AF880887

Euclid's Elements: Book I

A New Edition

HERBERT M. SAURO
SEATTLE

Ambrosius Publishing

First Edition, version 1.02
Published by Ambrosius Publishing
euclid.analogmachine.org

Typeset using LaTeX 2_ε, TikZ (3.1.10), WinEdt (10)
and 11pt Math Time Professional 2 Fonts

ISBN 13: 978-1-7325486-4-0 (paperback)

Printed at a location depending on purchase.

Mosaic image modified from Daniel Steger's Tikz image (http://www.texample.net/tikz/examples/mosaic-from-pompeii/).

Front-Cover: The front cover shows part of folio 43 from Heath manuscript designation: V. Titled: Vienna MS Philos Gr. No 103. This manuscript was digitized by the Austrian National Library, ONB Digital. The manuscript can be found at: https://digital.onb.ac.at/RepViewer/viewer.faces?doc=DTL_3360050 with designation: Opera cum Scholiis Euclides, ca. v3. Jh.; Busbecq, Ogier Ghislain, 1522-1592. Cod. Phil. gr. 31 HAN MAG. The manuscript appears to have changed designation since Heath published his book in 1925. Cod. Phil. gr. 31, however, matches exactly the description given by Heath (1925) on page 48 of his book, right down to the folio numbers (top right/left corners of the folios) including the content of the manuscript: Euclid books I-XV, Optics, Phenomena and Scholia.

EUCLID'S ELEMENTS
BOOK I

A new edition of Book I.

To Holly, Theodore and Tyler

Preface

Having been intrigued for many years by the scientific and mathematical works of the ancient Greeks, I decided to take the plunge in the summer of 2022 to thoroughly read at least Book I of Euclid's elements. In so doing I was inspired to write a new edition of Book I of the elements. To be different from all the other 1000+ published editions of Euclid, I decided to make it a hybrid of Byrne's colorful and purely pictorial edition of Euclid and the more traditional text based books.

This edition largely uses the original language of Euclid, though has been updated in places with less terse language. The text is based largely on the Greek translation by Fitzpatrick (2008), located at `https://github.com/rfitzp/Elements`, and the edition by Heath (1925), where the text can be obtained from the 1925 2nd edition at `archive.org/details/euclid_heath_2nd_ed/1_euclid_heath_2nd_ed` as well as the Perseus project which provides a more accurate version in text form at:

`http://www.perseus.tufts.edu/hopper/text?doc=urn:cts:greekLit:tlg1799.tlg001.perseus-eng1`

All the figures have been completely redrawn in color, with color markers in the main text for easier cross-referencing. I have added a number of sections at the beginning of the book that discuss the history of the text and other items of interest. I have added my own short commentaries on a number of the propositions, as well as an appendix that covers some of the prepositions in more detail. The ultimate authority on Euclid's Elements is, and continues to remain, Heath (1925).

This was a project to satisfy my own curiosity and I had no initial intention of publishing this booklet. But as the book grew, I though perhaps others might find the work of interest and given today's ease for publishing, I decided to issue the work as a small booklet. Applying the same model to the other 12 books of Euclid would be quite time consuming; and it might be something that I attempt upon retirement.

As always, I must thank my infinitely patient wife, Holly, for helping to proof read the text. All remaining errors are of course my own responsibility.

December, 2022.

Contents

1 Introduction

Euclid's Elements is one of the most famous books in the world. What may be surprising to learn is that the book is about mathematics, written 2,300 years ago. It is second only to the Bible in printed editions, with over 1000 different editions. The book lays out a method of deductive reasoning that has probably influenced western thought more than any other book. It continues to be republished even in the 21st century, with numerous blogs and other articles discussing its merits and content.

Euclid's Elements is a collection of 13 books that lay out the foundation for geometry, number theory, and many core concepts of mathematics and logic still important today. Its popularity is evident by the fact it was still used in schools and colleges well into the 20th century as a textbook on geometry. Eventually it was supplanted, though reluctantly, by more modern treatments. There is, however, continuing interest in using Euclid's Elements as a means to teach deductive reasoning, and as a valuable way of introducing the concept of proof in mathematics.

The entire collection of books is comprised of definitions, postulates, and a large number of mathematical proofs, many of which are related to geometric constructions. Such proofs are called propositions in Euclid's Elements. The 13 books cover plane and solid Euclidean geometry, elementary number theory, and incommensurable lines.

This publication will focus on Book I of the Euclid's Elements. Book I is the most well known of the thirteen books because it contains Proposition 47, the most famous theorem of all, Pythagoras' theorem:

> *In right-angled triangles the square on the side subtending the right angle is equal to the squares on the sides containing the right angle.*

Book I is about basic geometrical properties. For example, in Book I we get the theorems about congruent triangles, including: SAS (Prop 4), SSS (Prop 8) and ASA (Prop 26). We also get the theorem that tells us that in a triangle, the three interior angles always add to 180°. Amazingly, we still use the same proof today that Euclid described 2,300 years ago. Book I therefore contains a lot of basic theorems that are needed to do more advanced geometry.

1.1 A Brief History

The history of Euclid's Elements is quite complicated. It survived 2,300 years of wars, famine, and the general upheaval of civilizations as they rose and fell. A complete picture of how Euclid's Elements reached us is therefore still an active area of research. The following section gives a simplified and incomplete history of Euclid's Elements.

Ancient Times

Surprisingly, given how important the book is, very little is known about the author. Certainly Euclid was Greek and lived around 300 BC in Alexandria, Egypt. In writing Euclid's Elements, he brought together work from a variety of authors spanning the previous 200 years or more, as well as adding his own material.

After Euclid, we know of various commentaries, such as Geminus (~1st century BC), Heron (~60 A.D), Porphyry (~233-309 A.D), Pappus (~290-350 A.D), Proclus (412-485 A.D), and Simplicius (490-560 A.D.). This reveals that interest in Euclid was still very active during the period between 300 B.C and 500 A.D. Fragments of these commentaries have survived, including the complete commentaries from Proclus on Book I. The work by Proclus is particularly important because he refers back to earlier authors, whose original work has been lost. The commentaries offer an important glimpse into the issues that were of interest to these early mathematicians. Heath (1925, volume 1)[1] describes in some detail the work of these authors.

The primary text that has come down to us from this period (until the discovery of Vatican 190 in 1808) is the edition by Theon of Alexander from the 4th century AD (335-405 A.D). Theon, like Euclid, worked in Alexandria as a teacher of mathematics and astronomy. We know roughly when he lived because he says he observed a solar eclipse on 16 June 364. Theon was also the father of Hypatia, considered by some to be one of the last classical Greek scholars.

The original manuscript that Theon wrote didn't survive. Instead, a copy was made in Constantinople in 888 AD and by some rare chance, this copy survived and can still be seen today in the Bodleian library in Oxford. The manuscript has also been digitized so anyone can view the manuscript from the comfort of their own home[2].

In Figure 1, I've summarised how Euclid's Elements came down to us through the ages. This is a simplified timeline and there is much that isn't known. The figure also only highlights four main language translations, Greek, Arabic, Latin and English. Clearly

[1] When I refer to Heath (1925), I will always mean Volume 1 of that work. See also Dover edition 2012.
[2] `https://digital.bodleian.ox.ac.uk/objects/d4a23501-0b98-4aff-acd6-fe06fe9b62e3/`

there are editions in other languages such as French, German, Hebrew, Chinese, etc. which are not included.

Early Middle Ages

The first non-Greek name that has come down to us after Proclus is the Roman Boethius from the 4th/5th century. He appears to have created a Latin translation of Euclid's Elements, but very little remains other than fragments, and even the extent of the translation is unknown (Busard, 2004). However, evidence suggests that his writings influenced subsequent editions in 12th century Europe. Another important name is Al-Hajjaj (786–833 CE). He was a mathematician who lived in Baghdad. He was responsible for translating Euclid's Elements from Greek into Arabic, which eventually found its way into Spain in the 11th and 12th centuries, along with other Arabic editions. At that time, Spain was occupied by several Arabic Caliphates who supported learning and scholarship, including a great library at Cordoba. It was within this tolerant environment that individuals traveled from northern Europe to seek ancient Greek manuscripts such as Euclid's Elements.

Late Middle Ages

In the 12th century Adelard of Bath translated an Arabic copy to Latin based on his travels to southern Europe, the so-called Adelard I. Where he got this Arabic edition from is not known for certain. One older book claims he disguised himself as a Muslim student to obtain a copy in Muslim Córdoba[3], however the basis of this is no certain. It is more likely he either translated a manuscript while in Sicily, or may have had access to the material from somewhere like Chartres, France, or Antioch (in present day Turkey). Studies suggest (Busard, 2004) that this early copy by Adelard is entirely based on Arabic sources as no Greek or Latin influences, such as Boethius, have been detected in the text. This implies that these early translators did not have easy access to the original Greek manuscripts or even early Latin translations, as least in Spain and England. The whereabouts and influence of the Greek texts during the period from 888 A.D to 12th century is uncertain. Other translators at this time include Gherard (or Gerard) of Cremona, who translated a copy from Arabic to Latin; a Euclid manuscript

[3]Ball, Walter William Rouse (1915). A Short Account of the History of Mathematics (6th ed.). MacMillan.

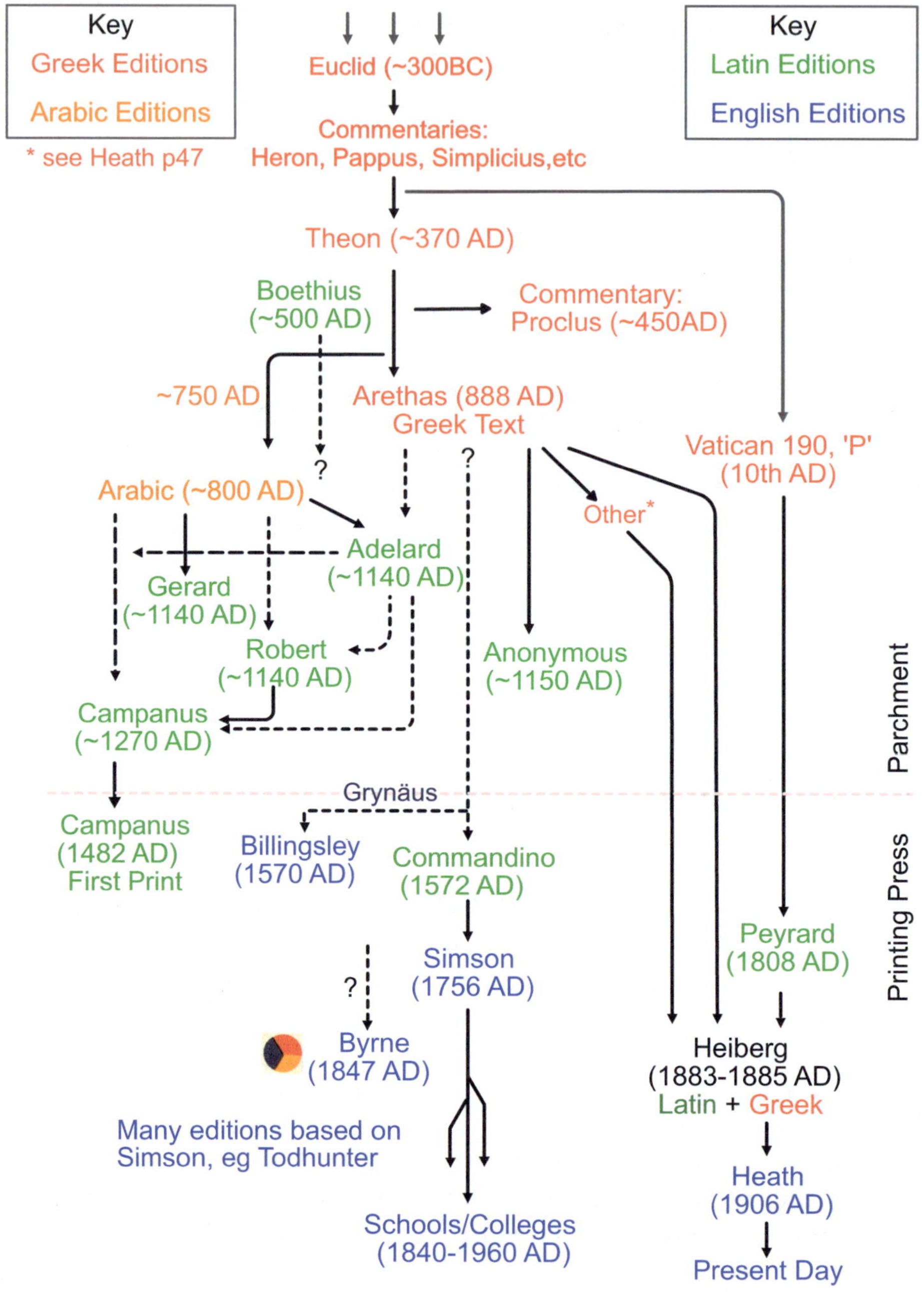

Figure 1: A simplified history of the Euclid's Elements through time. Solid lines mean certain, dotted lines mean uncertain, but probable. See also Figure 3.

by Gerard was discovered in 1901 by A, A Björnbo in the Vatican Reg. lat. 1268. Another important translator was Robert of Chester (traditionally known as Adelard II but actually written by Robert of Chester) who also made a translation from Arabic to Latin, but with some likely influence from Boethius (Busard, 2004). Hermann of Carinthia, who appears to have been a close friend of Robert of Chester when they were both in Spain, also made a translation to Latin from Arabic. The translation by Robert of Chester only contains summaries of the propositions. This edition however would later become the basis for subsequent editions in medieval Europe.

What is not so well known, however, is an anonymous translator, who translated a Greek version of Euclid into Latin around the same time. Apparently this translation was only discovered as recently as the 1960s in the Bibliothèque Nationale in Paris. This translation was done by a medical student from Salerno. Having heard that a copy of Ptolemy's Algemest was available in a library at Palermo, Sicily, he travelled to see this manuscript. Evidently, while at Palermo, our anonymous translator also came into contact with Euclid's Elements, where he made a Latin translation, possibly even from the Bodleian Theon manuscript or at least a Theon derived manuscript. However, this translation never seems to have had an impact on subsequent authors. The Latin manuscript is extant, available online and is located at the Bibliothèque Nationale in Paris, Fonds latin, 7373, 178 fols., ca 18x27 cm., vellum, 13th century (`https://gallica.bnf.fr/ark:/12148/btv1b10035274r/`). The translation to Latin is largely complete and covers Books I to XIII and XV with abbreviated versions of the non-Euclidean Books XIV and XV. The date for translation is thought to be somewhere between 1160 and 1180.

Interestingly, a second translation from the 14th century and almost identical to the Palermo edition was discovered in 1962 at the Firenz, Bibliotheca Nazionale Centrale, C I 488. This manuscript is only complete up to X.48 but contains the missing definitions, postulates, and common notions of Book I together with the lost enunciation of Proposition 1. The Palermo copy had lost these sections which had been replaced, at a later date, with Adelard I versions. Full accounts of these manuscripts can be found in an article by Murdoch, 1967, and a publication of the work in the book by Busard, 1987. Figure 2 show a fragment of page from the 12/13th century Greek to Latin translation from Salerno, showing Propositions 47.

In the 13th century (1255-1259), we see the translation of Euclid to Latin by the Italian Campanus of Novara. This edition was based on previous translations, particularly those by Robert of Chester, Adelard of Bath, and Gerard of Cremona. The Campanus edition

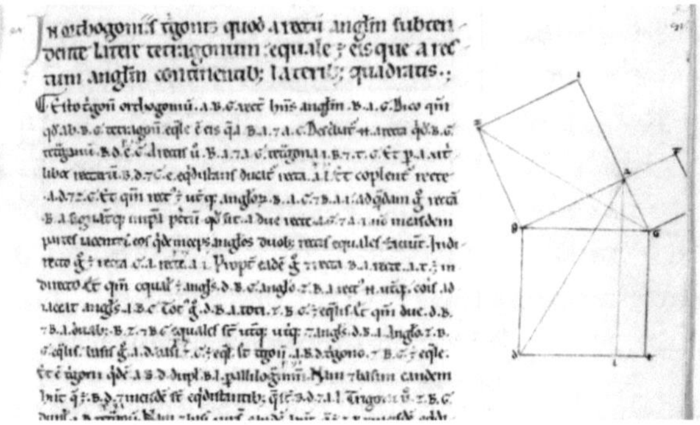

Figure 2: A fragment from the unique 12/13th century Greek to Latin translation from Palermo, Sicily. It shows part of Propositions 47, the infamous Pythagorean Theorem. Bibliothèque Nationale in Paris, Fonds Latin, 7373. The entire manuscript can be seen at `https://gallica.bnf.fr/ark:/12148/btv1b10035274r/`

would later become the first printed Latin edition of Euclid, published by Ratdolt at Venice in 1482.

It is evident that the survival of Euclid during early to mid medieval times is complex and still not fully understood, and continues to be an active area of research.

Renaissance

Before the invention of the printing press, manuscripts had to be laboriously copied by hand, which meant very few copies were made. This changed when, between 1436 and 1446, the German goldsmith Johannes Gutenberg is credited with inventing and developing the printing press. Whereas a scribe might take weeks to a year to make one manuscript, a printing press could print hundreds of books in a few weeks. From producing one copy manually to printing 100s, or even 1000s of copies, the printing press caused a profound change in how knowledge was disseminated.

By 1500, the printing presses in Western Europe had already produced more than twenty million books (Febvre and Martin, 1976) on every possible topic (today we pubish 4 million books a year). In the following century, output rose tenfold to an estimated 150 to 200 million copies. Compare this to the 1000s of handwritten manuscripts that had been produced in the previous 300 years.

With the invention of the printing press, the number of copies of Euclid in circulation exploded. By 1600, Euclid had been translated and printed into most European languages, including a new Greek translation to Latin by Zamberti, 1505. The Greek codex used by Zamberti is unfortunately not known (Riccardi, 1828), but appears to be a Theon based text. In 1533 Gryneaus published the first printed Greek edition apparently based on two Greek manuscripts (Venetus Marcianus 301 and Paris gr. 2343[4]) as well Zamberti's Latin edition (Heath, 1925, page 101). The Codex Marciana 301 is most likely located in the Library of Saint Mark, Venice (Italian: Biblioteca Marciana). The Venetian manuscript does not appear to be available online however, nor is there any information describing it other than a reference in a thesis by Eunsoo Lee[5] and Heath 1925, page 101. The Paris gr. 2343 manuscript is available on line, see footnote 2.

In 1572, Commandino published his Latin version of Euclid which was based on a different set of Greek manuscripts. At least one of these was based on the printed Greek manuscript by Grynaeus in 1533. The precise manuscripts Commandino used, however, is unknown, though it is suggested that he didn't use the Greek texts held in the Vatican (Davide Pietrini, 2020). Nevertheless, it was Commandino's edition that ultimately supplanted the original Campanus edition.

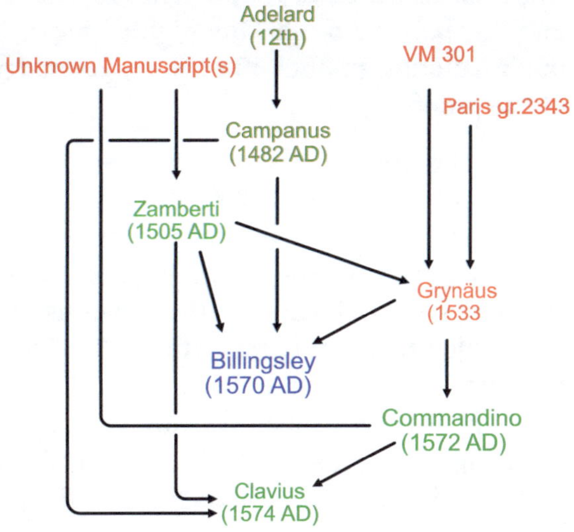

Figure 3: Simplified summary of the relationships between the main Euclid editions during the Renaissance. Red:Greek; Green:Latin; Blue:English

The Greek editions that Zamberti and Commandino used likely originated either from the Vatican library or from new copies that were brought over from Constantinople and other libraries in the east. New wealthy renaissance families such as the Medici family were avid collectors of manuscripts, including Greek manuscripts. The most famous of these was Lorenzo de Medici, who in the late 15th century, sent individuals to the east to search for manuscripts, including copies of Euclid (von Reumont, 1876). Clearly there were Greek

[4] `https://archivesetmanuscrits.bnf.fr/ark:/12148/cc224727`

[5] Visual agency in Euclid's elements : a study of the transmission of visual knowledge, 2920

versions of Euclid in circulation at this time, many it seems dating from the 10th-12th centuries (see the list in the section on early manuscripts). Wardhaugh mentions in his book 'Encounters with Euclid' that in 1808, Peyrard had access to 23 editions of Euclid[6] that were on loan from the major libraries in Italy, including the Vatican, as a result of Napoleon's conquest of most of Italy in 1799.

In the English speaking world, the first English translation of Euclid was by Billingsley in 1570 in London, with the title: "The elements of geometrie of the most auncient philosopher Evclide of Megara". Other than differences in spelling from today, you may also notice that Euclid is referred to as Euclid of Magara. Unfortunately this was the wrong Euclid. Euclid of Megara was a philosopher who lived around 400 BC. This was a frequent source of confusion during the Renaissance. Billingsley's copy was translated from the Greek edition by Gryneaus[7], as well as using information from the Campanus and Zamberti editions. One highly innovative aspect of Billingsley's book was the use of pop up models in Book XI on solid geometry, which gave readers a better understanding of the proofs.

The final edition that should be mentioned during this period is the edition by the German Jesuit Clavius in 1574[8]. As Clavius states himself, this wasn't a new translation, but rather a consolidation of exiting editions and commentaries. It was by far the longest and most complete edition of the Elements ever produced, and collected almost everything that had been published in the previous centuries. Some (De Risi, 2016) consider Clavius' edition to be more important than Commandino's because of its wider impact. Clavius based his work on at least the editions by Campanus, Zamberti, and Commandino[9]. His book was very successful and was reprinted and updated many times over the years, including 1574, 1589, 1591, 1603, 1607 and Frankfort 1612. After his death in 1612 further editions were published in 1617, 1627, 1629, 1639, 1654, and 1655 (Wardhaugh, et al., 2020). The book was originally published as a two volume set but later could also be purchased as a single three inch thick edition that included all the books. Clavius' edition is also famous because it eventually made its way to China, resulting in a Chinese translation of Euclid's Elements[10].

Figure 3 gives a simplified summary of the relationships between the main editions

[6] This matches information in the Dictionary of Greek and Roman Biography and Mythology (1870) - Volume 2, page 72

[7] See `https://www.maa.org` Billingsley's sources

[8] https://archive.org/details/euclidiselemento01eucl/page/n13/mode/2up

[9] http://www.sphere.univ-paris-diderot.fr/IMG/pdf/en_renaissanceeditionseuclidelements_caracteristics.pdf

[10] https://www.loc.gov/item/2021666487

published during the Renaissance.

Modern Era

With so many editions being published, it was inevitable that authors would try to make changes to the text to help distinguish themselves from other editors. This meant the text started to evolve away from its original form. This ultimately led to attempts to publish editions that were closer to the original text of Euclid.

In the English speaking world, a number of authors attempted to do this, but it was the Scottish mathematician, Robert Simson, who was the most successful. In the last page of the Preface, Simson stated, “Upon these Accounts it appeared necessary, and I hope will prove acceptable to all Lovers of Accurate Reasoning and of Mathematical Learning, to remove such blemishes, and restore the principal Books of the Elements to their original Accuracy.” Simson’s edition had set the stage for most subsequent editions in English and over 70 different editions, including translations, were made based on his book.

Copies, somewhat modified, were even being published as late as 1933 in the UK as part of the long running Todhunter series. Almost all school and college textbooks were based on Simson’s edition. If you’re old enough to have been taught geometry using Euclid in high school, then it’s an edition

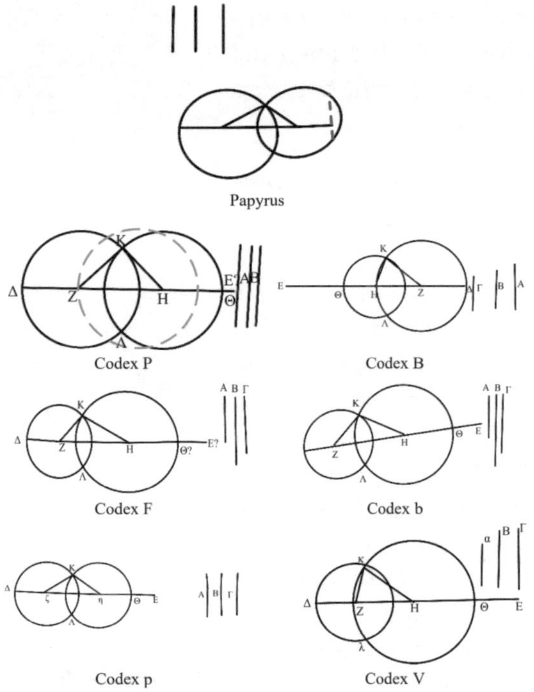

Figure 4: Different renderings of the figure associated with Proposition 22 from various manuscripts. Image taken from “Visual Agency in Euclid’s Elements: A Study of the Transmission of Visual knowledge” Eunsoo Lee, Stanford University ProQuest Dissertations Publishing, 2020, 28103932.

that can trace its lineage back to Simson, and from there back to Theon.

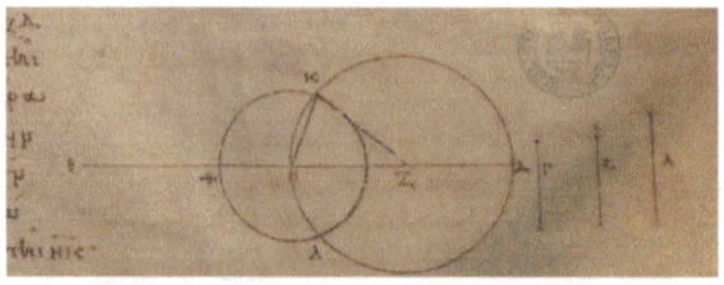
B - MS Bodleian Theon 888AD, Heath p47

F - MS XXVIII.3, Heath p47

P - MS Vatican 190, Heath p47

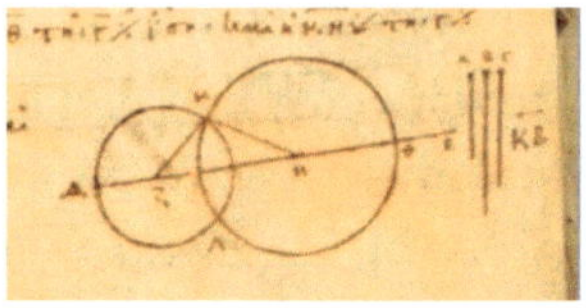
b - MS Bologna, Heath p49

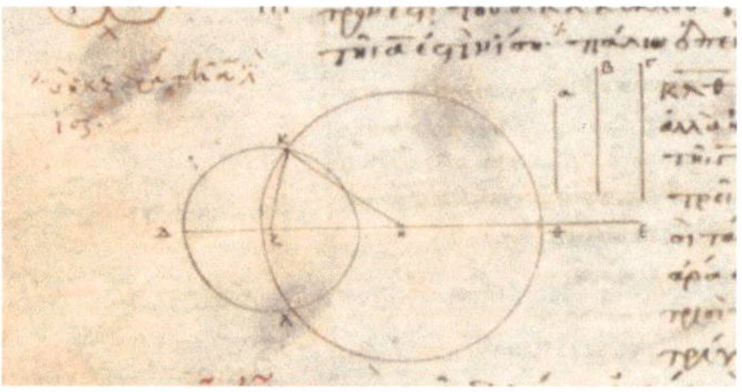
V - MS Vienna, Heath p48

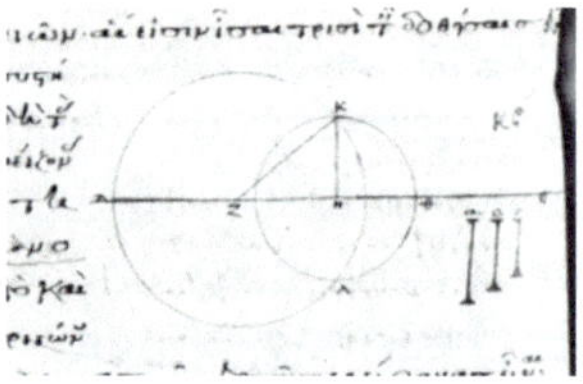
q - MS Paris (2344), Heath p50

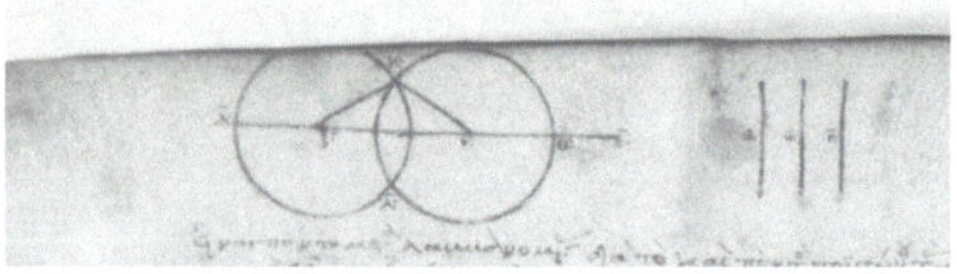
p - MS Paris (2466), Heath p49

Figure 5: These are actual images of the figures from Proposition 22 from various manuscripts including q - Paris 2466 which is not included in the previous figure. Some are symmetrical (P, p), while others are mirror images (V, q). The reference to Heath indicates the page number that describes the corresponding manuscript in Heath 1925.

Most translations and editions of the Elements have been based on Theon's original version of Euclid. What is surprising however is how varied the texts are. Not only is the prose different in places, but the number of postulates and axioms differ quite markedly. Even the figures are different between editions.

One can easily spot differences in the figures by comparing the online version of the 888 AD edition with later published versions. For example, the figure associated with Proposition 22 shows a variety of different renderings (Figure 4). This figure shows seven renderings from seven different manuscripts (also called codexes). Figure 5 shows screenshots from the actual manuscripts.

You will also notice that each subfigure of Figure 4 has a Codex letter, such as P or V. These letters come from Heath who described many early editions of Euclid's Elements and used simple letters to indicate which manuscript he was referring to. Thus 'P' is the Vatican 190 edition from the 10th century.

Manuscript P

At the start of the 19th century, all was about change.

In 1808, Peyrard discovered a new 10th century version of Euclid from the Vatican. Known as manuscript P, or Vatican 190, this manuscript was older, and thus pre-Theon. In other words, it was likely closer to the original Euclid that Theon's edition. Heiberg, in the 1870s, carried out a careful study of many extant manuscripts, including the newly discovered P and published a new edition of Euclid in 1883.

This edition is thought to be closer to the original than any other published version. The Heiberg edition formed the basis for Heath's famous and extremely thorough analysis of Euclid's Elements. Heath's version has since been reedited and published more recently (2002-2013) by Green Books. Richard Fitzpatrick (2008) also recently published a completely new translation from Heiberg's Greek text.

The two variants, the Heiberg and the Simson editions, can be easily distinguished by the number of postulates set out in Book I. Simson and all editions derived from Simson have only three postulates, whereas the Heiberg derived editions have five. The other stark difference about Simsons' edition is that it has twelve axioms, while Heiberg's has just five. Even more surprising, an inspection of the Clavius edition of 1574 shows 19 axioms. Throughout the history of Euclid, there has been significant variation in the number of postulates and axioms. De Risi, 2016 describes in detail the great variety of postulates and axioms used in the various editions.

Finally, it would be amiss not to mention the famous edition of Euclid by Byrne. In 1847, Oliver Byrne published a most unusual edition of Euclid that dispensed with the text and replaced it with colorful diagrams. Each proposition is described using coloured shapes, lines and angles. You can see an example shown under Proposition 5 later in this book.

1.2 Early Manuscripts

Many of the earliest manuscripts such as the Theon 888 AD edition, or the famous pre-Theon P manuscript can be found online in digital format. In fact, all the manuscripts mentioned by Heath (1925) in Chapter V, pages 46 to 50 are now available online. Many of these likely formed the basis for the new Greek and Latin translations in the 15/16th centuries.

For those interested, I've put the list of web links to the manuscripts discussed by Heath below.

1. Heath Designation: P
Title: Theon Vatican 190 (pre-Theon)
Date: 10th (Greek)
Link: `https://digi.vatlib.it/view/MSS_Vat.gr.190.pt.1`

2. Heath Designation: F
Title: XXVIII.3 Laurentian Library, Florence
Date: 11th (Greek) - Heath suggests 10th but the library catalog states 11th
The manuscript contains Euclid's Elements in the first part of the book.
Link: `http://mss.bmlonline.it/s.aspx?Id=AWOIR02hI1A4r7GxMFba&c=III#/book`

3. Heath Designation: B
Title: Theon Vatican 190
Date: 888 AD (Greek)
Link: `https://digi.vatlib.it/view/MSS_Vat.gr.190.pt.1`

4. Heath Designation: V
Title: Vienna MS Philos Gr. No 103 (Now designated Cod. Phil. gr. 31 HAN MAG)
Date: $\approx$ 12th, library catalog claims Constantinople, 1100
The manuscript contains Euclid's Elements in the first part of the book.
Link: `https://digital.onb.ac.at/RepViewer/viewer.faces?doc=DTL_3360050&order=1`

5. Heath Designation: b
Title: Bologna Numbered 18-90 two volumes
Date: 11th Heiberg conjectures that this was written by a Byzantine mathematician.
Euclid's Elements start at folio 95
Link: `http://badigit.comune.bologna.it/books/A18/scorri.asp`

6. Heath Designation: p
Title: Paris MS. 2466
Date: 12th (Greek)
Link: `https://gallica.bnf.fr/ark:/12148/btv1b107229634`

7. Heath Designation: q
Title: Paris MS. 2344
Date: 12th (Greek)
Link: `https://gallica.bnf.fr/ark:/12148/btv1b10723139w#`

1.3 Current Editions

Of the current editions of Euclid's Elements, I am aware of three paperbound English versions:

1. The first is an excellent rerendering of Heath's edition by Dana Densmore and is therefore considered close to the original pre-Theon edition. This includes all thirteen books in one binding. This edition is quite popular and has been reprinted numerous times (with corrections) since it was originally published in 2002.

2. The second edition to consider is by Richard Fitpatrick. This version uses a two column format, with the left column containing the Greek version from Heiberg's Greek edition and the right column is the author's own English translation of Heiberg's Greek text. Heiberg's edition is pre-Theon. Fitpatrick's edition is available as a pdf online but also has a print edition available on Amazon. Please note there are two versions, one from 2007 which does not use the two-column format so that the Greek and English translations have their own dedicated pages. The 2008 version follows the formatting as seen in the pdf version. The version of the text used in this book follows closely, but not completely, the Fitpatrick translation.

3. The third edition is by Kathryn L. Goulding. This is an interesting edition because one of its aims is to use Euclid as a teaching aid. The edition was first published in 2004 but appears to have a 2017 reprinting. It covers Book I to VI of Euclid. As with the other books, it's available on Amazon for a very modest sum. It comes in two versions, Instructor and Student. What I liked about the Instructor's edition is that at the start of the book, the author discusses the problems and recommendations for teaching Euclid to today's audience. Goulding based her text on the web site edition by David Joyce (`http://aleph0.clarku.edu/~djoyce/elements/elements.html`). This in turn is based on a slightly modified version of the Heath edition. As with all Heath/Heiberg texts, they are pre-Theon.

There is also a teachers Guide to Euclid by J.J Jacob Tawney (2020).

There are a number of web sites that cover Euclid's elements. The most well known is David Joyce's web version which is based on an adaption of Heath's edition. Joyce includes a large number of commentaries and is well worth looking at. Others include Pythagoreanmath[11], Bookofproofs[12], and a really nice site dedicated to Byrne's edition[13]. This is not an exhaustive list.

There are of course editions published in previous decades such as the Great Books of the Western world volume 10, published by Robert P Gwinn in 1952, and reprinted in 1990. Many earlier Euclid books can be purchased on the second-hand market. Alternatively, you can go to Archive.org where you'll find umpteen editions and copies of Euclid from the 21st, 20th and 19th centuries.

Finally, it is worth mentioning YouTube - if only the Greeks had access to YouTube.

[11] `https://pythagoreanmath.com/`

[12] `https://www.bookofproofs.org/branches/euclids-elements/`

[13] `https://www.c82.net/euclid/`,

There are a number of excellent channels, and I list four here. Euler's Academy[14], mathematicsonline[15], Mathoma[16] and last but not least, Sandy Bultena[17] who covers every book from Euclid's Elements. Of those listed, my favorites are Sandy Bultena and Euler's Academy channels.

It is remarkable that after 2,300 years, Euclid's Elements is still very much alive and well.

Heath's Edition of Euclid

I'd like to specifically highlight Heath's edition, 1925 (currently published by Dover, 2012), which has been mentioned previously many times. Heath's edition is a translation from Heiberg's Greek texts, and many of the modern editions follow the work of Heath.

Published in two editions, 1906 and 1925, these books are packed with useful information. Due to the enormous number of notes Heath makes on every definition, postulate, axiom and proposition, the book is printed in three separate volumes.

Volume 1 covers Book I and II, Volume 2 covers Book III to IX, and Volume 2 covers books X to XIII. Anyone with an interest in Euclid's Elements should have these copies on their shelf. At the very least, readers should have the first volume which contains a wealth of analysis and historical notes. The written style may be considered old fashioned by some, but I personally found it easy to read. As a piece of scholarship, it is unsurpassed. This level of scholarship is rarely seen today due to how academic life is now structured via funding rounds and short-term goals.

1.4 This Edition

Many of the editions published over the last 100 years have strived to reproduce what is thought to be close to the original text. This is, of course, of great historical interest. However, the original text, although readable, could be expressed in more modern English. The text is also very terse in places which can make it difficult for the novice.

Many of the editions published between the 15th and 18th century tried to update the text, and some authors added even more text to the propositions (in some cases significantly more) in their attempt to help explain the proofs. As a result, many of these early

[14] https://www.youtube.com/watch?v=RxzAH2qfT5U&list=PL2V76rȧjvC1I2TrbPMRLcTqhdcbha4sDE

[15] https://www.youtube.com/playlist?list=PLFC65BA76F7142E9D

[16] https://www.youtube.com/watch?v=g28xAeoARME&list=PLpzmRsG7u_gqSOYQWOPfJrCCQL-gfYfnw

[17] https://www.youtube.com/@SandyBultena

Author	Year	First Line
Adelard	12th	Now we have to demonstrate how the assigned line is divided into two halves.
Billingsley	1570	To divide a right line given being finite, into two equal parts.
Simson	1762	To bisect a given finite straight line, that is, to divide it into two equal parts.
Heiberg	1883	To divide the given straight line into two equal parts.
Todhunter	1903	To bisect a given finite straight line, that is, to divide it into two equal parts.
Heath	1925	To bisect a given finite straight line.
Joyce	1998	To bisect a given finite straight line.
Fitzpatrick	2008	To cut a given finite straight-line in half.
Densmore	2013	To bisect a given finite straight line.
Goulding	2017	To bisect a given finite straight line.

Table 2: Proposition 10 enunciation. Fitzpatrick is a new translation from Heiberg's Greek edition and as you can see it resembles Google's attempt at translating the Latin equivalent from Heiberg's text. Note that the Todhunter editions have changed over time, the earlier ones are slightly different from the 1903 edition shown here. The Adelard title is based on Google's translation of Adelard's Latin edition. Note: Since the original publication of this book, I would now recommend using ChatGPT to do Latin/Greek translations to English.

editions drifted away from the original text. The danger is that the flow of the proof or subtle nuances in the proof are lost. From the 18th century, however, there were efforts to recover as much of the original text as possible, culminating, as mentioned before, with the work by Heiberg, who in 1883, published editions in both Greek and Latin.

To give you some idea of the variability in the text, Table 2 shows the enunciation for Proposition 10. One can see that in the last 100 years the text has been quite stable, but between the 12th and 20th century, it was highly variable.

The following blocks show Proposition 10 for various editions from the 12th century to the current time. Note the degree of variability between editions. Adelard from Bath is particularly interesting as it looks almost completely different from any other, as well as being much shorter. This likely reflects the fact that Adelard derived his edition from an Arabic rather than a Greek source.

Adelard of Bath 12th century from Latin

Now we have to demonstrate how the assigned line is divided into two halves.

Let BA be the assigned line, and let a triangle of equal sides be ***AGB*** on it.

Therefore, since we have shown that the line ***GA*** is equal to the line GB, and the line ***GD*** is common, the lines ***AG*** and ***GD*** will be reciprocally equal to the lines ***BG*** and ***GD***

Henry Billingsley 1570

To divide a right line given being finite, into two equal parts.

Suppose that the right line given be ***AB***, It is required to divide the line ***AB*** into two equal parts. Describe (by the first proposition) vp. on the line ***AB*** an equilateral triangle and let the same be ***ABC***.

And (by the former proposition) divide the angle ***ACB*** into two equal parts by the right line ***CD***.

Then I say that the right line ***AB*** is divided into two equal parts in the point ***D***.

For for as much as (by the first proposition) ***AC*** is equal to ***CB***, and ***CD*** is common the both, therefore these two lines ***AC*** and ***CD*** are equal to these two lines ***BC*** and ***CD***, y one to y other, and the angle ***ACD*** is equal to the angle ***BCD***.

Wherefore (by the 4. proposition) the base ***AD*** is equal to the base ***BD***.

Wherefore the right line given ***AB***, is divided into two equal parts in the point ***D*** which was required to be done.

Robert Simson 1762

To bisect a given finite straight line, that is, to divide it into two equal parts.

Let ***AB*** be the given straight line; it is required to divide it into iwo equal parts.

Describe upon it an equilateral triangle ***ABC***, and bisect angle ***ACB*** by the straight line ***CD***, ***AB*** is cut into two equal parts in the point ***D***.

Because ***AC*** is equal to ***CB***, and ***CD*** common to the two triangles ***ACD***, ***BCD***; the two sides ***AC***, ***CD*** are equal to ***BC***, ***CD***, each to each; and the angle ***ACD*** is equal to the angle ***BCD***; therefore the base ***AD*** is equal to the base ***DB***, and the straight line ***AB*** is divided into two equal parts in the point ***D***. Which was to be done.

Isaac TodHunter 1903

To bisect a given finite straight line, that is, to divide it into two equal parts.

Let ***AB*** be the given straight line:

it is required to divide it into two equal parts.

Construction. On ***AB*** describe an equilateral triangle ***ABC*** [I. 1.]

and bisect the angle ***ACB*** by the straight line ***CD***, which meets ***AB*** at ***D***. [I. 9]

AB shall be cut into two equal parts at the point ***D***.

Proof. In the triangles ***ACD***, ***BCD***,

because ***AC*** is equal to ***CB*** [Def. 23] and ***CD*** is common to both, and the angle ***ACD*** is equal to angle ***BCD*** [Construction]

therefore the base ***AD*** is equal to the base ***DB*** [I. 4.]

Wherefore the given straight line ***AB*** is divided into two equal parts at the point ***D***. [Q.E.F.]

Johan Heiberg from the Latin 1883

To divide the given straight line into two equal parts. Let the given straight line be ***AB***.

it is therefore necessary to divide the bounded straight line ***AB*** into two equal parts. Let an equilateral triangle $AB\Gamma$ be constructed in it [prop.1] and the angle.

$A\Gamma B$ the line $\Gamma\Delta$ is divided into two ae-like parts [prop. IX]

I mean, the straight line ***AB*** is divided at the point Δ into two equal parts.

For when $A\Gamma = \Gamma B$, and $\Gamma\Delta$ is common, the two straight lines $A\Gamma, \Gamma\Delta$ the two $B\Gamma, \Gamma\Delta$ are equal to each other;

and angle $A\Gamma\Delta = B\Gamma\Delta$. why/how $A\Delta = B\Delta$ [prop. IV]

Therefore, the given straight line AB at the point $\triangle$ is divided into two equal parts;

Thomas Heath 1926/1956/2012

To bisect a given finite straight line.

Let ***AB*** be the given finite straight line.

Thus it is required to bisect the finite straight line ***AB***.

Let the equilateral triangle ***ABC*** be constructed on it, and let the angle ***ACB*** be bisected by the straight line ***CD***;

I say that the straight line ***AB*** has been bisected at the point ***D***.

For, since ***AC*** is equal to ***CB***, and ***CD*** is common, the two sides ***AC***, ***CD*** are equal to the two sides ***BC***, ***CD*** respectively;

and the angle ***ACD*** is equal to the angle ***BCD***; therefore the base ***AD*** is equal to the base ***BD***. [I. 4]

Therefore the given finite straight line ***AB*** has been bisected at ***D***.

Richard Fitzpatrick 2008

To cut a given finite straight-line in half.

Let ***AB*** be the given finite straight-line. So it is required to cut the finite straight-line ***AB*** in half.

Let the equilateral triangle ***ABC*** have been constructed upon ***AB*** [Prop. 1.1],

and let the angle ***ACB*** have been cut in half by the straight-line ***CD*** [Prop. 1.9].

I say that the straight-line ***AB*** has been cut in half at point ***D***.

For since ***AC*** is equal to ***CB***, and ***CD*** (is) common,

the two (straight-lines) ***AC***, ***CD*** are equal to the two (straight-lines) ***BC***, ***CD***, respectively.

And the angle ***ACD*** is equal to the angle ***BCD***. Thus, the base ***AD*** is equal to the base ***BD*** [Prop. 1.4].

Thus, the given finite straight-line ***AB*** has been cut in half at (point) ***D***. (Which is) the very thing it was required to do.

Dana Densmore 2013

To bisect a given finite straight line.
Let ***AB*** be the given finite straight line.
Thus it is required to bisect the finite straight line ***AB***.

Let the equilateral triangle ***ABC*** be constructed on it, and let the angle ***ACB*** be bisected by the straight line ***CD***;

I say that the straight line ***AB*** has been bisected at the point ***D***.

For, since ***AC*** is equal to ***CB***, and ***CD*** is common, the two sides ***AC***, ***CD*** are equal to the two sides ***BC***, ***CD*** respectively;

and the angle ***ACD*** is equal to the angle ***BCD***;

therefore the base ***AD*** is equal to the base ***BD***. [1. 4]

Therefore the given finite straight line ***AB*** has been bisected at ***D***. Q.E.F.

Kathryn Goulding 2017

To bisect a given finite straight line.

Let ***AB*** be the given finite straight line.

It is required to bisect the finite straight line ***AB***,

Construct the equilateral triangle ***ABC*** on it, and bisect the angle ***ACB*** by the straight line ***CD***.[I.1, I.9]

I say that the straight line ***AB*** is bisected at the point ***D***.

Since ***CA*** equals ***CB***, and ***CD*** is common, therefore the two sides CA and CD equal the two sides ***CB***

and ***CD*** [I.Def 20, I.4] respectively, and the angle ***ACD*** equals the angle ***BCD***, [I.Def 20]

therefore the base ***AD*** equals the base ***BD***.

Therefore the given straight line ***AB*** is bisected at ***D***. Q.E.F

2 Summary of Book I

Book I is about basic geometry, setting up the initial definitions and axioms followed by 49 propositions. The propositions are the core of the book and build upon each other. Most of the propositions prove a result about geometry or a geometric construction based on the axioms and postulates, but also on prior propositions. For example, the first proposition explains how to draw an equilateral triangle, i.e a triangle will three identical sides. Proposition 2 shows how one can duplicate a line to another place on the drawing surface. However, in doing so, Proposition 2 has to be able to draw an equilateral triangle. Hence Proposition 2 depends on Proposition 1. This happens throughout the book, making the order of propositions very important. Figure 6 illustrates the dependencies (based on work by Joyce) between the different propositions in Book I.

There are two types of proposition:

- **Constructions or demonstrations**, also called problems in early editions. These describe, as the name suggests, the construction of some geometric problem. Examples include Proposition 1 which shows how to construct an equilateral triangle.

Constructions always end with the abbreviation Q.E.F which stands for *quod erat faciendum* or in English, "precisely what was required to be done".

- **Theorems**, these describe a truth about some aspect of a geometric shape. An example includes Proposition 5 which proves certain relationships about angles in an isosceles triangle. Theorems always end with the abbreviation Q.E.D which stands for *quod erat demonstrandum* or in English, "precisely what was required to be proved".

Each proposition has a statement on its purpose, but because the list can be a bit daunting at first, I've grouped related propositions together as shown in the following list. One interesting observation is that Group 5 contains a single entry, Proposition 15. In more modern texts this preposition is usually found near the beginning of the text book, sometimes even as the first proposition (For example Schultze, 1925, Plane Geometry).

Group 1: Equilateral triangles, basic manipulation theorems, and SAS theorem.

1: Construction of an equilateral triangle.

2: Construction: Duplicating a straight-line.

3: Construction: Removing a given part from a straight-line.

4: Triangle side-angle-aide (SAS) equality.

Group 2: Isosceles Triangles

5: Isosceles triangle has two equal angles.

6: Triangle with two equal angles is isosceles (converse of previous one).

Group 3: SSS Theorem

7: Two lines meet at a unique point.

8: Triangle side-side-side (SSS) equality.

Group 4: Bisection and Perpendicular Theorems

9. Bisection of angle.
10. Bisection of straight-line.
11. Construction of perpendicular line.
12. Perpendicular through given point.
13. Two angles on straight-line make two right angles.
14. Two angles making two right angles make straight-line.

Group 5: Opposite Angles

15. Two straight lines make equal opposite angles.

Group 6: Properties of Triangles

16. External angle of triangle greater than internal opposite.
17. Two angles of triangle less than two right angles.
18. Greater side of triangle subtends greater angle.
19. Greater angle of triangle subtended by greater side.
20. Sum of two sides of triangle greater than third side.
21. Lines through endpoints of one side of triangle to point inside triangle is less than sum of other sides.
22. Construction of triangle from given lengths.
23. Construction of equal angle (i.e copying a triangle).
24. Hinge theorem.
25. Converse hinge theorem.
26. Triangle angle-side-angle (ASA) and side-angle-angle (SAA) equality.

Group 7: Parallel Lines

27. Equal alternate angles implies parallel lines.
28. Equal corresponding angles or supplementary interior angles implies parallel lines.
29. Parallelism implies equal alternate angles, corresponding angles, and supplementary interior angles.
30. Parallelism is transitive relation.
31. Construction of a parallel line.

Group 8: Angles of a Triangle

32. Part 1: External angle of triangle equals sum of other internal angles.
    Part 2: Sum of angles of triangle equals two right angles.

Group 9: Parallelogram and Triangle Theorems Related to Areas

33. Lines joining equal and parallel straight-lines are parallel.
34. Opposite sides and angles of parallelogram are equal.
35. Parallelograms with same base and same height have equal area.
36. Parallelograms with equal base and same height have equal area.
37. Triangles with Same base and same height have equal area.
38. Triangles with equal base and same height have equal area.
39. Equal sized triangles on same base have same height.
40. Equal sized triangles on equal base have same height.

Group 10: Area Arithmetic Propositions

41. Parallelogram on same base as triangle has twice its area.
42. Construction of parallelogram equal to triangle in given angle.
43. Complements of parallelograms are equal.
44. Construction of parallelogram on given line equal to triangle in given angle.
45. Construction of parallelogram in given angle equal to given polygon.
46. Construction of square on given straight-line.

Group 11: Pythagoras' Theorem

47. Pythagoras's Theorem.
48. Square equals sum of squares implies right triangle.

Propositions 45 and 47 represent the pinnacle of Book I.

3 The Other Books

All together, there are thirteen books in Euclid's Elements. As we've discussed, Book I is concerned with triangles, parallels and areas. The table below lists the contents of the other books. Historically, the books were often split into two, Book I to VI, then Books VII to XIII for more advanced students. This is how the Clavius 1574 two-volume edition was split.

In total, considering all 13 books as described by Heath, 1925, there are 465 propositions, 131 definitions, 5 postulates, and 5 common notions or axioms. The number of common notions however depends on the edition, for example Simson editions have 12.

BOOK	Contents
BOOK I	Triangles, parallels, and area
BOOK II	Geometric algebra, including construction of the golden ratio
BOOK III	Circles and their properties, including Thales' Theorem
BOOK IV	Constructions for inscribed and circumscribed figures, including regular polygons with 4, 5, 6, and 15 sides
BOOK V	Theory of proportions
BOOK VI	Similar figures and proportions
BOOK VII	Fundamentals of number theory including prime numbers
BOOK VIII	Continued proportions in number theory
BOOK IX	Number theory including the infinitude of prime numbers
BOOK X	Classification of incommensurables, including the irrationality of the square roots
BOOK XI	Solid geometry
BOOK XII	Measurement of figures including volumes of cones, pyramids, and cylinders
BOOK XIII	Regular solids

A summary of how many items in each Book can be found in Table 3[18]. This data is based on the edition by Heath, 1925.

4 Notation used in this edition

I've taken the liberty to color the figures in the propositions, which can be exploited by adding corresponding colored notation to the text. This is partly inspired by Oliver Byrne's famous edition of Euclid, where Byrne created a truly novel edition that used color and pictures with minimal text and virtually no labels on the diagrams. Both lines and angles were color coded using four colors: red, blue, yellow and black. To the colors he also added solid and dotted lines. New editions of Byrne's Euclid can now be purchased as hard copy as well as being available online. If one is ever stuck with understanding a Proposition, I have found it worthwhile to go over Byrne's version as it can sometimes help. As an example, I have recreated Proposition 5 using Byrne's notation on page 45, so that you can compare it against Euclid's original prose. Anyone

[18]Data from `https://www.obkb.com/dcljr/euclid_orig.html`

Book	Definitions	Postulates	Axioms	Propositions
I	23	5	5	48
II	2	-	-	14
III	11	-	-	37
IV	7	-	-	16
V	18	-	-	25
VI	3	-	-	33
VII	22	-	-	39
VIII	-	-	-	27
IX	-	-	-	36
X	16	-	-	115
XI	28	-	-	39
XII	-	-	-	18
XIII	-	-	-	18
Totals	120	5	5	465

Table 3: Number of items in each Book.

wishing to purchase a second-hand copy of Byrne's book at their local second-hard book store may, however, be disappointed. First-edition copies are currently priced at around $20,000.

In this edition, angles in a proposition will always be indicated with a colored symbol such as ∠. This is in addition to the normal Euclidian notation that uses lettering. The color of the symbol will correspond to the coloring of the matching angle in the figure. For example:

The angle ***CDA*** (∠) is equal to the angle ***ABC*** (∠).

Sometimes two angles are combined, as in (∠ + ∠).

Straight edges will be indicated with a colored line, for example (—). I won't always do this, because sometimes it simply adds unnecessary clutter to the text. At times however, the eye is quicker at finding a particular color then trying to spot the labeling in the figure. An example of a possible sentence that uses both notations is:

Let the angle ***ABC*** (∠) be cut in half by the straight-line CD (—).

Sometimes a line segment may be composed of two or more segments; these will be indicated using this form of notation (——).

Circles, triangles and parallelograms are also indicated by colored symbols, for example (●), (▲) and (■). The particular color is used to identify the object in question according to the colors used in the figure.

5 Further Readings

The following are recommend books, articles, and websites (should they still exist) for the interested reader. Many of the books can be bought on Amazon for a modest price although the research books tend to be more expensive. The list also includes references made in the main text.

H.L.L Busard (2005) Campanus of Novara and Euclid's elements, Steiner, Stuttgart.

H.L.L. Busard, The Mediaeval Latin Translation of Euclid's, Elements (1987), Stuttgart, Franz Steiner Verlag Wiesbaden GmbH ,411 pp.

De Risi, V. The development of Euclidean axiomatics. Arch. Hist. Exact Sci. 70, 591–676 (2016). https://doi.org/10.1007/s00407-015-0173-9

Dana Densmore (2002-2013) Euclid's Elements, all thirteen books complete in one volume. Green Lion Press.

Febvre, Lucien; Martin, Henri-Jean (1976). The Coming of the Book: The Impact of Printing 1450–1800. London: New Left Books. Quoted in: Anderson, Benedict. Comunidades Imaginadas. Reflexiones sobre el origen y la difusión del nacionalismo. Fondo de cultura económica, Mexico, 1993. ISBN 978-968-16-3867-2. pp. 58f.

Richard Fitpatrick (2008) Euclid's Elements of Geometry (English edition): edited, and provided with a modern English translation.

David Fowler, The Mathematics of Plato's Academy, 2d edition, Oxford, 1999.

Katheryn Goulding (2004-2017) Euclid's Elements Book I-VI With Exercises (Instructor and Student editions).

Thomas Heath (1925) Euclid's Elements Books I to II. (reprinted by Dover in 1956/2012).

Thomas Heath (1925) Euclid's Elements Books III to IX. (reprinted by Dover in 1956/2012).

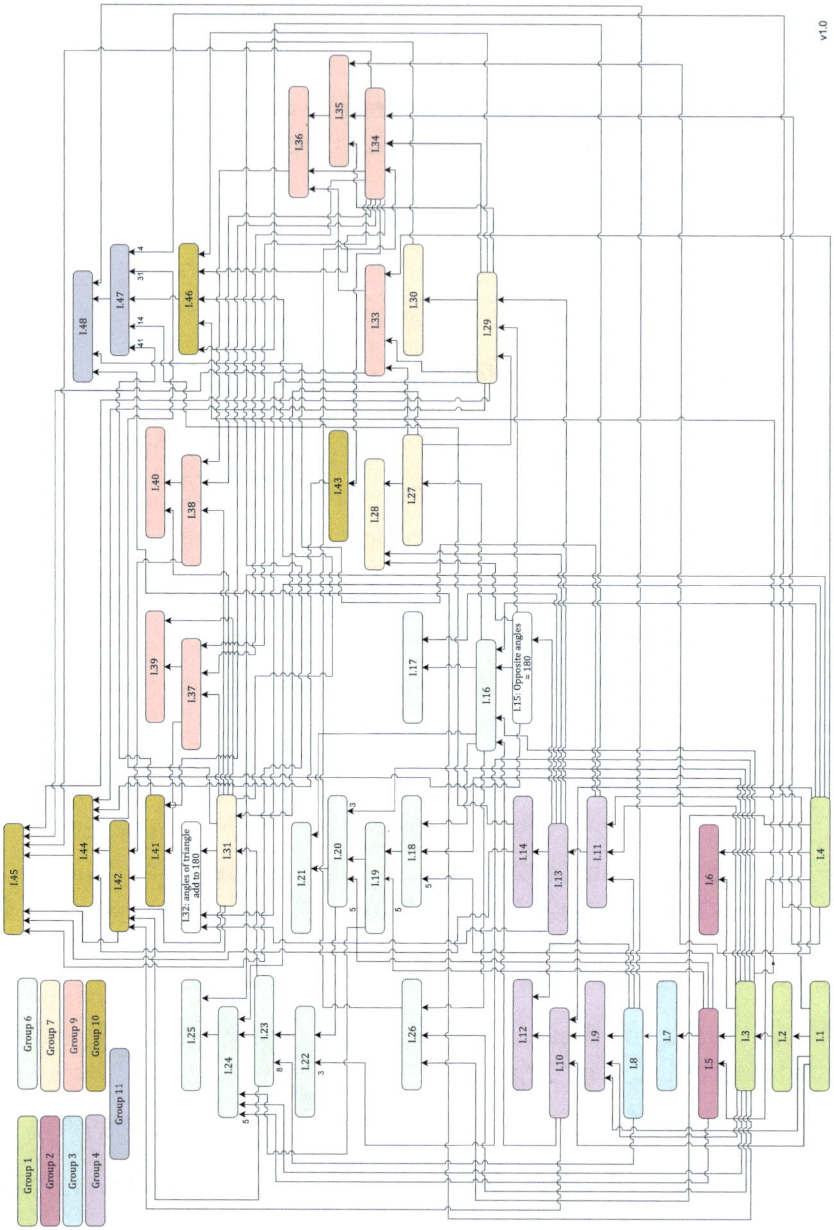

Figure 6: Proposition dependencies in Book I. Based on image by `https://www.reddit.com/user/graphbang/` but redrawn as a pdf, errors corrected and rearrangements made. Data from David Joyce site. Available at `https://github.com/hsauro/euclid`

Thomas Heath (1925) Euclid's Elements Books X to XIII. (reprinted by Dover in 1956/2012).

David E Joyce (1996-present time), Euclid's Elements, Books I to XIII. Web site: http://aleph0.clarku.edu/~djoyce/elements/elements.html

Murdoch, John Emery. "Euclides Graeco-Latinus: A Hitherto Unknown Medieval Latin Translation of the Elements Made Directly from the Greek." Harvard Studies in Classical Philology 71 (1967): 249.

Davide Pietrini (2020) University of Urbino, PhD Thesis, pages 31-32 https://ora.uniurb.it/retrieve/handle/11576/2683371/133727/phd_uniurb_285800.pdf

Riccardi, Pietro (1887) Saggio di una bibliografia euclidea, Page 5, paragraph 2 (Roughly translated "Essay from a Euclidean bibliography"), See archive.org for copy.

Alfred von Reumont (1876) Lorenzo De' Medici, the Magnificent (vol 1 of 2).

Ken Saito Diagrams and traces of oral teaching in Euclid's Elements: labels and references. ZDM Mathematics Education 50, 921–936 (2018). https://doi.org/10.1007/s11858-018-0929-1

J. Jacob Tawney (2020) Problems and Proofs: A Teacher's Guide to Euclid's Elements Spiral-bound by Cana Academy. ISBN-13: 978-1734056785

Wardhaugh, Benjamin, Philip Beeley, and Yelda Nasifoglu. Euclid in Print, 1482–1703: A Catalogue of the Editions of the Elements and Other Euclidean Works. Reading Euclid project, University of Oxford, 2020.

Jensen Newell (2023) A new version of Oliver Byrne's book that has been completely recreated using LaTex and figures from https://www.c82.net/euclid/ by Nicholas Rougeux.

6 Supplemental Material

The repository at https://github.com/hsauro/euclid will store supplementary material. Currently this site holds a pdf copy of the dependency figure (Figure 6). This material is freely available for readers to use so long as an appropriate attribution is provided.

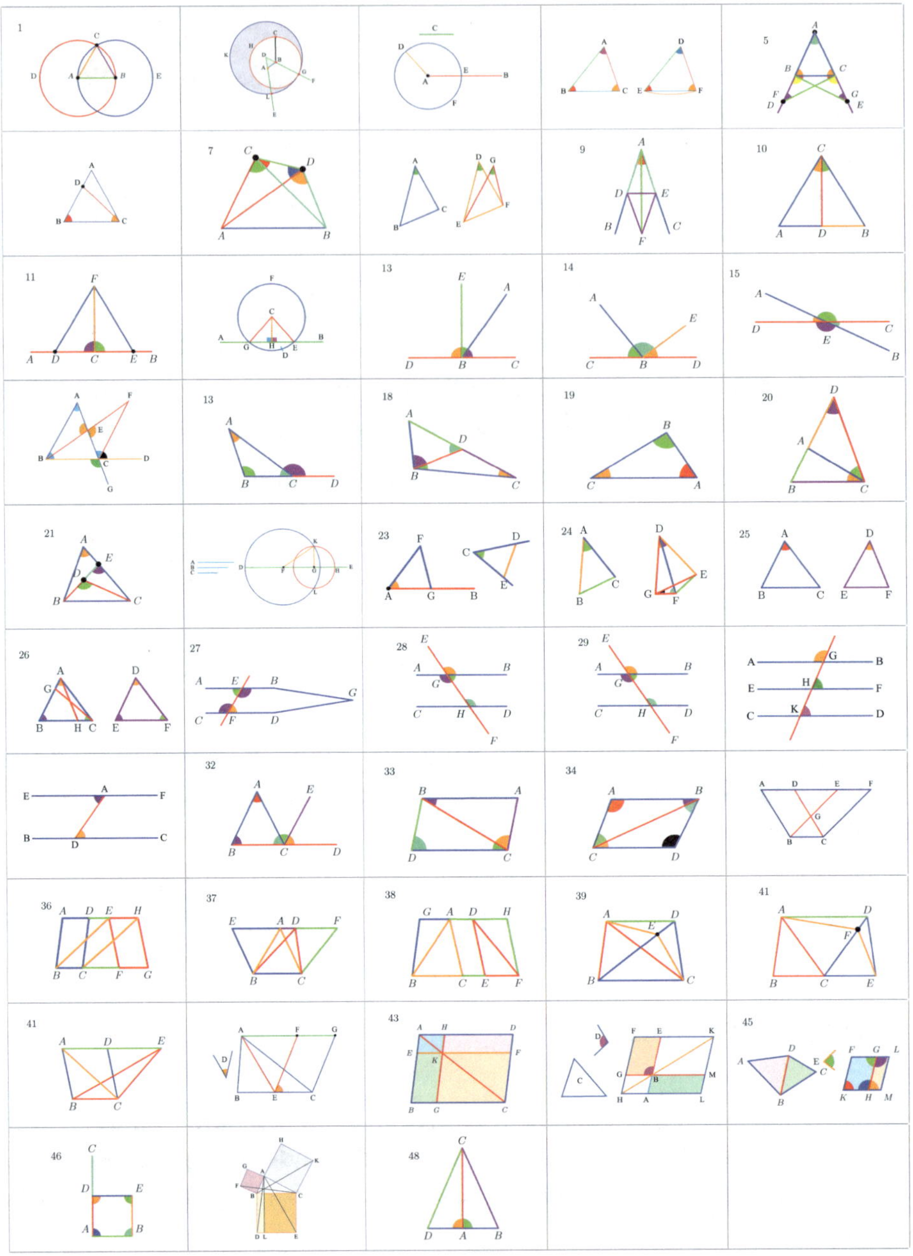

Figure 7: Matrix of all figures in Book I.

EUCLID'S ELEMENTS
BOOK I
Elementary Plane Geometry

DEFINITIONS

1. A **point** is that which has no part.

> **Note:** That is, a point is simply a location in space, a zero dimensional object. This is often indicated by drawing a small filled circle at the location.

2. A **line** is a length without breadth.

> **Note:** This is a harder one to explain since a line made with a pencil does indeed have a breadth (or width). However, what is being indicated here is an idealized line that allows us to connect two points. A line under this definition can be a straight or curved line.

3. The **extremities** of a line are points.

4. A **straight-line** is one which lies evenly with points on itself.

> **Note:** This is an attempt to define what a straight line is. The definition is somewhat vague and modern geometry does not try to define a straight line this way. Instead a 'line' is the name of something that has specific properties. For example, one property is that one and only one straight line can be drawn between any two points.

5. A **surface** is that which has length and breadth only.

6. The extremities of a surface are lines.

7. A **plane surface** is surface which lies evenly with the straight-lines on itself.

8. A **plane angle** A, is the inclination of the lines to one another, when two lines in a plane meet one another, and are not lying in a straight-line.

> **Note:** This may seem a duplicate of the next definition but it is more general in that it defines the angle between lines (not necessarily straight) where they meet.

9. When the lines containing the angle are straight, the angle is called **rectilinear**.

> **Note:** This the definition of an angle that we may be more familiar with, two non-parallel straight lines intersect at a point, the angle is the inclination between the two lines.

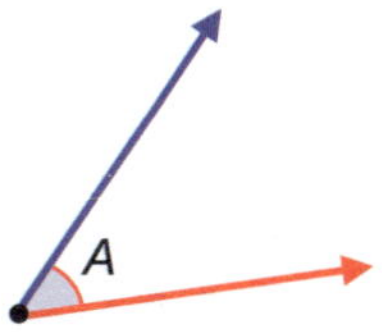

10. When a straight line set up on a straight line makes the adjacent angles equal to one another, each of the equal angles is **right**, and the straight line standing on the other is called a **perpendicular** to that on which it stands.

> **Note:** This is a wordy definition for a perpendicular line. Today we would just say that a perpendicular line is one that intersects another line at 90 degrees.

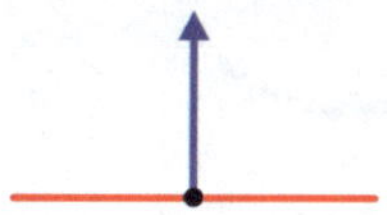

11. An **obtuse angle** is one greater than a right-angle.

12. And an **acute angle** is less than a right-angle.

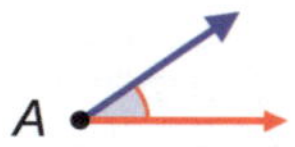

13. A **boundary** is that which is the extremity of something.

14. A **figure** is that which is contained by some boundary or boundaries.

15. A **circle** is a plane figure contained by a single line called a **circumference**, such that all of the straight-lines radiating towards the circumference from one point amongst those lying inside the figure are equal to one another.

> **Note:** Note that the radiating lines represent the radius of the circle.

16. The point, from which the equal lines are drawn, is called the **center** of the circle.

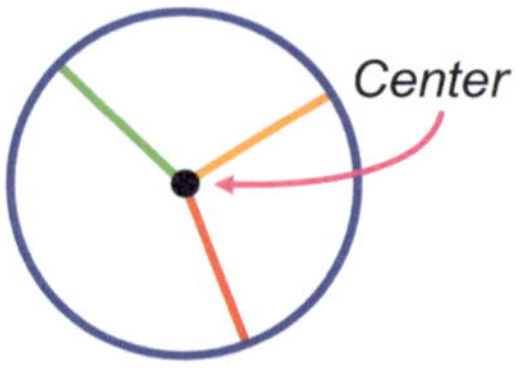

17. The **diameter** of the circle is any straight-line, being drawn through the center, and terminated in each direction by the circumference of the circle. Any such straight-line also cuts the circle in half.

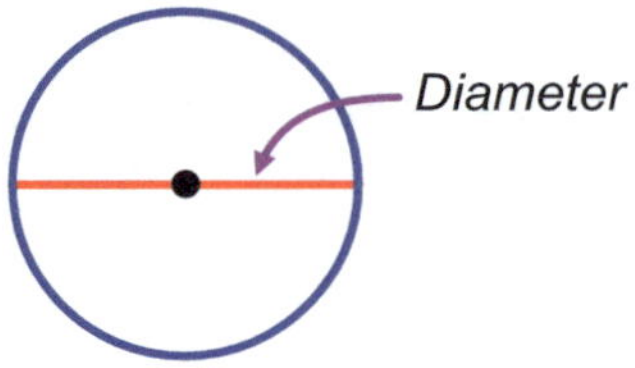

18. The a semi-circle is the figure contained by the diameter and the circumference cuts off by it. The center of the semi-circle is the same point as the center of the circle.

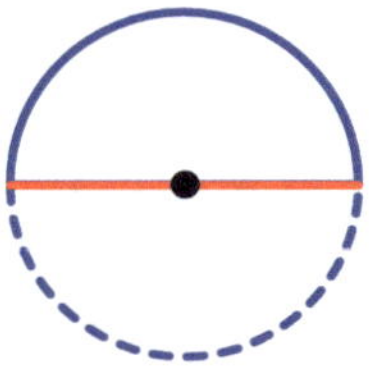

19. Rectilinear figures are those figures contained by straight-lines: **trilateral** figures being those contained by three straight-lines, **quadrilateral** by four, and **multilateral** by more than four.

20. Of the trilateral figures: an **equilateral triangle** is that having three equal sides, an **isosceles triangle** that having only two equal sides, and a **scalene triangle** that having three unequal sides.

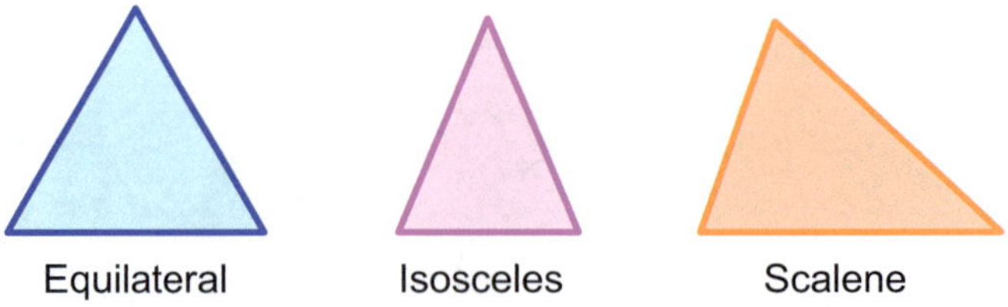

21. Further of the trilateral figures: a **right-angled triangle** is that having a right-angle, an **obtuse-angled triangle** that having an obtuse angle, and an **acute-angled triangle** that having three acute angles.

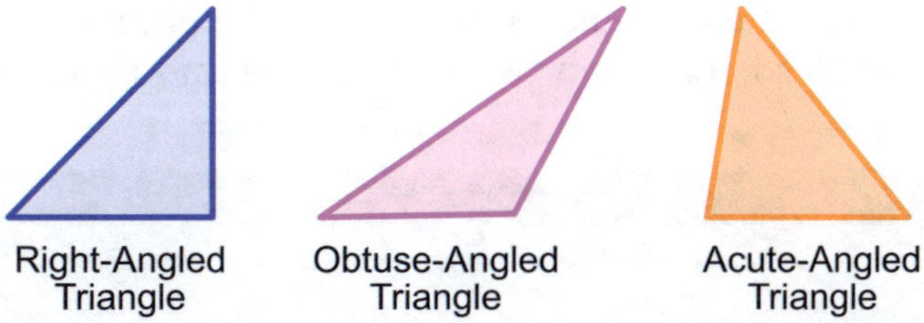

22. Of the quadrilateral figures: a **square** is that which is right-angled and equilateral, a **rectangle** that which is right-angled but not equilateral, a **rhombus** that which is equilateral but not right-angled, and a **rhomboid** that having opposite sides and angles equal to one another which is neither right-angled nor equilateral. And let quadrilateral figures besides these be called **trapezia**.

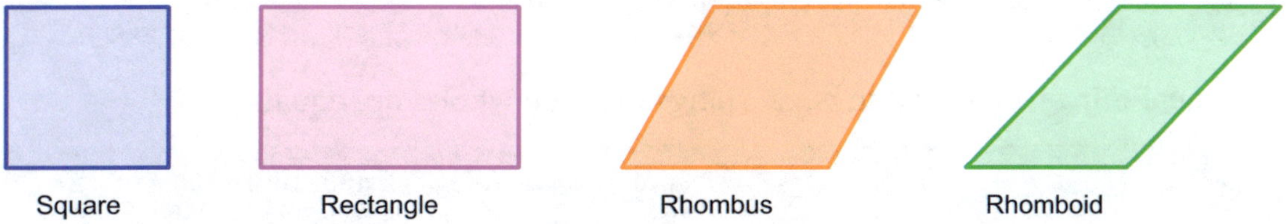

23. **Parallel lines** are straight-lines which, being in the same plane, and being produced to infinity in each direction, meet with one another in neither (of these directions).

POSTULATES

1. That a straight line may be drawn from any one point to any other point

2. That a terminated straight line may be produced to any length in a straight line

3. That a circle may be described with any centre, at any distance from that centre, that is, with any given line drawn from the centre as radius.

4. All right-angles are equal to one another.

5. If a straight-line falling across two other straight-lines makes internal angles on the same side of itself whose sum is less than two right-angles, then the two other straight-lines, being produced to infinity, meet on that side of the original straight-line that the sum of the internal angles is less than two right-angles and do not meet on the other side.

> **Note:** This is the infamous parallel postulate.

COMMON NOTIONS OR AXIOMS

1. Things equal to the same thing are also equal to one another.

> **Note:** That is, if two things are the same as something else then the two things must be the same.
> For example, we have three books, two of the books are the same, and the second book is the same as the third, therefore the first book must be the same as the third book.
> In modern notation we would write: if $A = B$ and $B = C$, then $A = C$.

2. If equal things are added to equal things then the wholes are equal.

> **Note:** This might seem an oddly worded sentence but what its saying is this. Let's assume we have two lines that are equal, e.g ***A*** and ***B***, and another line, say ***C***. What it says is if we add the line ***C*** to lines ***A*** and ***B***, then the new lines will still be equal, that is $A + C = B + C$.
> Here is another example, if we have two beakers containing 20 mls of water and we add to each beaker another 15 mls of water, then each beaker will contain 35 mls of water.

3. If equal things are subtracted from equal things then the remainders are equal.

> **Note:** This is similar to the last one but it subtracts rather than adds. That is if $A = B$, then $A - C = B - C$.
> To use the water example, if we have two beakers containing 20 mls of water and we remove from each beaker 5 mls of water, then each beaker will contain 15 mls of water.

4. Things coinciding with one another are equal to one another.

Note: Let's say we have two squares, ***A*** and ***B***, and we move one of the squares onto the other one. If both squares match up exactly, then we can say both squares are the same.

5. The whole is greater than the part.

Note: A good example to explain this is to take a round cake and cut out a slice. I think we can all agree that the slice will be smaller than the whole cake. In other words, if we start with an object ***A***, and break it up into parts, then the parts will inevitably be smaller than the object ***A***.

As you can see the axioms describe very basic properties which can't be easily proved from other simpler properties. Axioms are also often referred to as self-evident truths.

Additional Notes

Note: It is worth pointing out the difference between postulates and common notions. At first sight they seem to be concerned with the same thing. However, postulates are only related to geometrical matters whereas the common notions are general principles that aren't limited to geometry.

Since the time of Aristotle, there have been innumerable comments written on the validity and utility of the definitions, postulates and axioms. Both Heath (1925) and Proclus have extensive summaries of many of the issues that have been raised over the last 2,000 years. Part of the problem stems from the vagueness of some of the definitions such as point, line and straight line, the fact that some of the definitions are not even used, and whether the list of postulates and axioms are even complete (Simson has twelve axioms, while Heiberg has five). For example, there are a number of subtle assumptions used in some of the propositions (e.g 1 and 22), assumptions which are not justified by the existing set of definitions, postulates or axioms.

Possibly the biggest issue of all, and one that resulted in much discussion and gnashing of teeth was postulate five concerning parallel lines. From a long time it was thought that the fifth postulate could be proved using the other four postulates but this turned out not to be possible. We now know that the parallel postulate must be included as a postulate to develop Euclidean geometry and can't be proved. Without the parallel postulate, alterative, but equally consistent geometries are possible, the so-called non-Euclidian geometries.

Modern geometry has completely revised the foundations of Euclidean geometry to avoid any ambiguity, and hidden assumptions and to make sure the foundation is complete and logically consistent. Even though Euclid's Elements has its issues (relatively minor in the grand scheme of things I think), all the geometric properties proven in the subsequent propositions are still valid and the work itself remains an outstanding achievement. It has stood the test of time for over two thousand years and the theorems it describes and proves are as true today as they were when first proposed.

Proposition 1

On a given finite straight-line, construct an equilateral triangle.

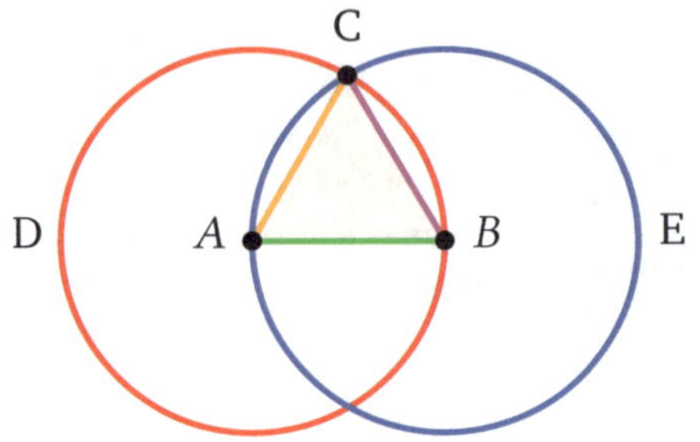

Let ***AB*** (——) be a finite straight-line.

Let us construct an equilateral triangle on the straight-line ***AB***.

Draw the circle ***BCD*** (●) with center ***A*** and radius ***AB*** (——) ***[Post. 3]***.

Draw the circle ***ACE*** (●) with center ***B*** and radius ***BA*** (——) ***[Post. 3]***.

Identify the point ***C*** where the circles cut each other. Draw lines ***CA*** (——) and ***CB*** (——) ***[Post. 1]***.

Since the point ***A*** is the center of the circle ***BCD*** (●), ***AC*** (——) is equal to ***AB*** (——) ***[Def. 15]***.

Since the point ***B*** is the center of the circle ***ACE*** (●), ***BC*** (——) is equal to ***BA*** (——) ***[Def. 15]***.

But ***CA*** (——) was also shown to be equal to ***AB*** (——). Thus, ***CA*** (——) and ***CB*** (——) are each equal to ***AB*** (——). But things equal to the same thing are also equal to one another **[Axiom 1]**.

Thus, ***CA*** (——) is also equal to ***CB*** (——). Thus, the three straight-lines ***CA***, ***AB***, and ***BC*** are equal to one another.

Therefore, the triangle ***ABC*** is equilateral, and has been constructed on the given finite straight-line ***AB*** (——). Q.E.F

Naming: The naming of the circles in Proposition 1 is inconsistent across editions and the ordering of the lettering, BCD and ACE, changes halfway through the proof. In fact, the order of lettering in many of the propositions seem to be quite fluid.

Criticisms of Proposition 1: Since the days of Zeno of Sidon (150–75 BC) there have been criticisms of Euclid's Elements (See Proclus). The first proposition in particular has received more than its fair share of critiques. The chief issue that has been brought up repeatedly is the assumption by Euclid that the two circles will in fact intersect each other at point *C*. Even if we assume the circles do intersect, how many intersection points will there be? The last question is actually answered in Book 3 Proposition 10 (A circle does not cut a circle at more than two points) but that's a bit too late. The first issue, that point *C* exists, is more serious however as there is nothing in the definitions or postulates that can help us. A possible way out of this is to first assume that a circle is a continuous line which encloses space. If each circle passes through the center of the other, then each circle must have a part of its circumference within the other, and a part outside it. Therefore there must be an intersection. This reasoning however requires an additional postulate or axiom called the circle-circle intersection axiom. This basically says "If one of two circles has one point inside the other and one point outside it, the two circles will meet." Heath claims to have a proof of this axiom based on work by Dedekind (Heath, 1925, p236). We should perhaps not be too harsh of a piece of work that is 2,300 years old. In 1899 Hilbert proposed a new axiomatization of geometry[a] that resolved many, if not all, the issues that had been raised in Euclid's Elements.

[a] `https://en.wikipedia.org/wiki/Hilbert%27s_axioms`

A note on the next three propositions: If this is the first time coming to Euclid, you might find the next two propositions, 2 and 3 a little puzzling and the fourth one perhaps out of place. Proposition 2 is about duplicating a line segment and 3 about cutting off segments. The reader is probably expecting to dive straight into proofs related to angles, triangles, and rectangles, much like a modern textbook on geometry. With Proposition 2, it might seem more convenient to just copy a line using a compass to physically measure a length, then move the compass to the new location and draw the copied line. However, the Greeks didn't like to use the compass-distance measure as part of a proof. Instead they preferred a purely geometric method to copy a line.

Proposition 2 is only referenced once in the whole of Book I where we find it used to prove Proposition 3. Proposition 3 in turn is used in twelve propositions in Book I as well as propositions in the other books on geometry in the series. Collectively Proposition 2 and 3 allow a given line to be copied to an another location in any orientation. For example, this technique is used in Proposition 5 to prove properties of angles. Moving to Proposition 4 we see it's a statement of the Side-angle-side (SAS) property but it's debatable whether it's an actual proposition. In fact modern treatments consider the SAS property to be a postulate. In view of these, you might think of Proposition 5 as the real starting point, with Propositions 1 to 4 as prerequisites.

In the first three propositions we also get a hint of how Euclid's Elements works. Each proposition is a new result that depends on previous proven propositions so that by the time we get to Pythagoras' theorem in Proposition 47, we know, based on the axioms, postulates, definitions and a host of earlier propositions, that the result must be true.

Proposition 2

To place at a given point, a straight-line equal to a given straight-line.

(1) Let ***A*** (orange) be the given point, and ***BC*** (—) the given straight-line. It is required to place a straight-line at point ***A*** equal to the given straight-line ***BC*** (—).

(2) Let a straight-line ***AB*** (—) be joined from point ***A*** to point ***B*** ***[Post. 1]***.

(3) Let an equilateral triangle ***DAB*** (▲) be constructed on this line ***[Prop. 1]***.

(4) Let the lines ***AE*** (—) and ***BF*** (—) be produced in a straight-line with ***DA*** (—) and ***DB*** (—) ***[Post. 2]***.

(5) Let a circle ***CGH*** (●) with center ***B*** and radius ***BC*** (—) be drawn ***[Post. 3]***.

(6) Let a circle ***GKL*** (●) with center ***D*** and radius ***DG*** (——) be drawn ***[Post. 3]***.

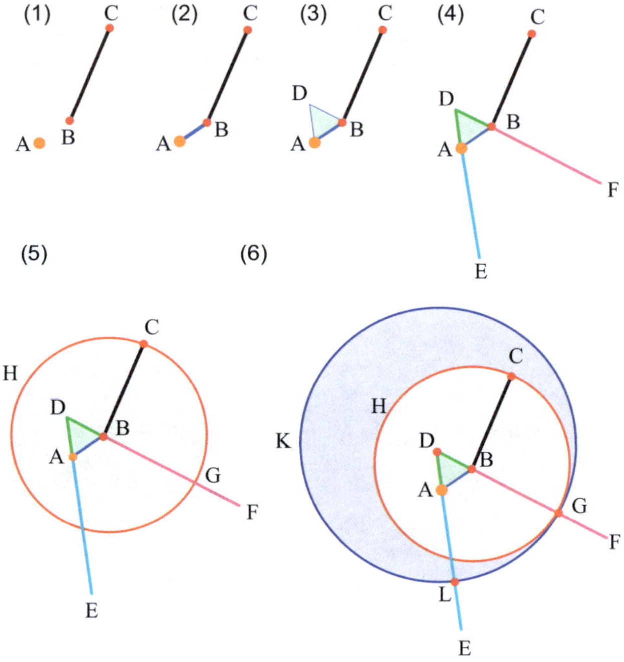

Since the point ***B*** is the center of the circle ***CGH*** (●), ***BC*** (—) is equal to ***BG*** (—) ***[Def. 15]***

Since the point ***D*** is the center of the circle ***GKL*** (●), ***DL*** (——) is equal to ***DG*** (——) ***[Def. 15]***.

Within these, ***DA*** (—) is equal to ***DB*** (—). Thus, the remainder ***AL*** (—) is equal to the remainder ***BG*** (—) **[Axiom 3]**.

But ***BC*** (—) was also shown to be equal to ***BG*** (—). Thus, ***AL*** (—) and ***BC*** (—) are each equal to ***BG*** (—). But things equal to the same thing are also equal to one another **[Axiom 1]**. Thus, ***AL*** (—) is also equal to ***BC*** (—).

Thus, the straight-line ***AL*** (—), equal to the given straight-line ***BC*** (—), has been placed at the given point ***A***. Q.E.F.

Proposition 3

Given two unequal straight-lines to cut off from the greater a straight-line equal to the lesser.

Let ***AB*** (——) and ***C*** (——) be the two given unequal straight-lines, where the greater be ***AB*** (——).

It is required to cut off a straight-line equal to the lesser ***C*** (——) from the greater ***AB*** (——).

Let the line ***AD*** (——), equal to the straight-line ***C*** (——), be placed at point ***A*** ***[Post. 2]***.

Let the circle ***DEF*** (●) be drawn with center ***A*** and radius ***AD*** (——) ***[Post. 3]***.

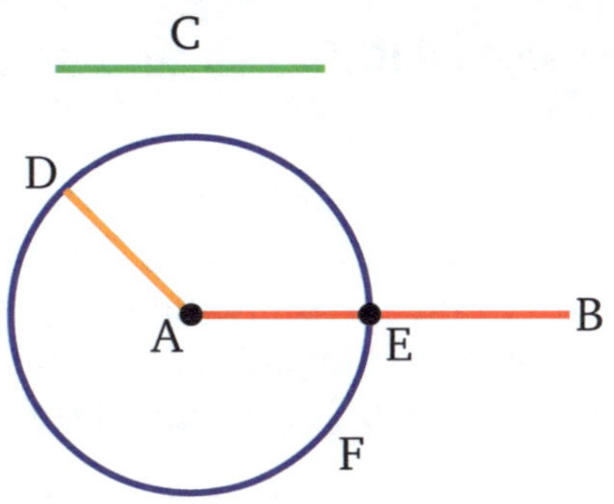

Since point ***A*** is the center of circle ***DEF*** (●), ***AE*** (——) is equal to ***AD*** (——) ***[Def. 15]***.

But, ***C*** is also equal to ***AD*** (——). Thus, ***AE*** (——) and ***C*** are each equal to ***AD*** (——). So ***AE*** (——) is also equal to ***C*** (——) **[Axiom 1]**.

Thus, for two given unequal straight-lines, ***AB*** (——) and ***C*** (——), the straight-line ***AE*** (——), equal to the lesser ***C*** (——), has been cut off from the greater ***AB*** (——).

Q.E.F.

Note: This proposition relies on Proposition 2 to move the line ***C*** to point ***A*** and is the only time Proposition 2 is used. Proposition 3 however is used twelve times. Appendix 8.1 has more commentary of these two propositions.

It is interesting to note that the figure used in Proposition 3 has two orientations in the literature. The one shown is the most common in editions from the last 500/1000 years, which includes the edition made by Adelard from Bath. I'll call this orientation 'horizontal'. Earlier manuscripts, particularly it seems those derived from Greek sources (such as P, the 888 AD Theon edition, and the anonymous manuscript from Palermo), draw the figure in the vertical orientation. This suggests that the vertical orientation might be the original diagram used by Euclid and for whatever reason has changed orientation in the intervening time.

Proposition 4

If two triangles have two sides equal to two sides respectively, and have the angles contained by the equal straight-lines equal, then they will also have the base equal to the base, and the triangle will equals the triangle, and the remaining angles will be equal to the remaining angles, namely those opposite the equal sides.

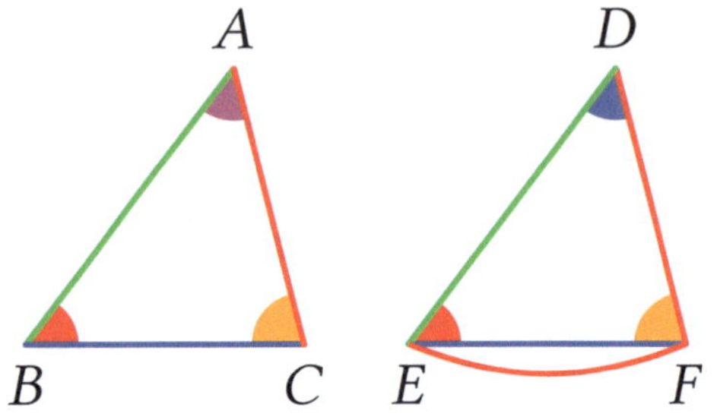

Let ***ABC*** and ***DEF*** be two triangles having the two sides ***AB*** (——) and ***AC*** (——) equal to the two sides ***DE*** (——) and ***DF*** (——), respectively. That is ***AB*** (——) to ***DE*** (——), and ***AC*** (——) to ***DF*** (——).

Let the angle ***BAC*** (∡) be equal to the angle ***EDF*** (∡).

I say that the base ***BC*** (——) also equals the base ***EF*** (——), the triangle ***ABC*** equals the triangle ***DEF***, and the remaining angles equal the remaining angles respectively, namely those opposite the equal sides, that is, the angle ***ABC*** (∡) equals the angle ***DEF*** (∡), and the angle ***ACB*** (∡) equals the angle ***DFE*** (∡).

If triangle ***ABC*** is applied to triangle ***DEF***, the point ***A*** being placed on the point ***D***, and the straight-line ***AB*** (——) on ***DE*** (——), then the point ***B*** will also coincide with ***E***, because ***AB*** (——) is equal to ***DE*** (——).

Because of ***AB*** (——) coinciding with ***DE*** (——), the straight-line ***AC*** (——) will also coincide with ***DF*** (——), on account of the angle ***BAC*** (∡) being equal to ***EDF*** (∡). So the point ***C*** will also coincide with the point ***F***, because ***AC*** (——) is again equal to ***DF*** (——).

But, point ***B*** coincides with point ***E***, so that the base ***BC*** (——) will coincide with the base ***EF*** (——) and equals it.

If ***B*** coincides with ***E***, and ***C*** with ***F***, and the base ***BC*** (——) does not coincide with ***EF*** (——), then two straight-lines will encompass an area: which is impossible.

Thus, the base ***BC*** (——) will coincide with ***EF*** (——), and will be equal to it **[Axiom 4]**.

So the whole triangle ***ABC*** will coincide with the whole triangle ***DEF***, and will be equal to it **[Axiom 4]**.

The remaining angles will coincide with the remaining angles, and will be equal to them **[Axiom 4]**.

That is ***ABC*** (∠) to ***DEF*** (∠), and ***ACB*** (∠) to ***DFE*** (∠).

Thus, if two triangles have two sides equal to two sides, respectively, and have the angles enclosed by the equal straight-line equal, then they will also have the base equal to the base, and the triangle will be equal to the triangle, and the remaining angles subtended by the equal sides will be equal to the corresponding remaining angles. Q.E.D.

Note: This is probably the weakest of all the proposition in Book I. It's a statement of the SAS rule (side-angle-side) for congruent triangles which many high School students should be familiar with. Some authors have felt that this proposition would have been best classified as a axiom, since it is fairly self-evident. In fact Hilbert, who recast Euclid in his publication 1899, defined proposition 4 as an axiom of congruence.

One thing to note about ***[Prop. 4]*** is that it doesn't depend on any of the previous three. It only depends on Axiom 4. Additionally, the labeling of dependencies is not consistent between editions. For example, Heath only indicates one Axiom 4 dependency while Densmore gives three. Simson doesn't indicate Axiom 4 at all and instead uses Axiom 10, which of course the original Euclid didn't even have that axiom. In fact Simson has 12 axioms in total compared to the five we normally see.

Curious about whether dependencies were indicated in older manuscripts, I checked both the Latin version from Adelard of Bath (translated from Arabic) and the anonymous Greek to Latin translation (12th century) from Pelermo (from Busard, 1987), and found no such indications of written dependencies. It would appear that using labels for the dependencies is a relatively modern addition. I have since came across this article[a] which comes to the same conclusion, to quote "...Euclid never gave explicit references in his proofs, although he did sometimes allude to previous propositions". The article also points out that more modern editions chose different dependencies and there is little consistency. See Appendix 8.3.

Of all the editions I have seen I would probably tend to go with the web edition by David E. Joyce when it come to listing dependencies. These appear more complete than others. The dependencies map in Figure 6 is based on Joyce's work.

[a] https://www.maa.org/book/export/html/590371

Proposition 5

In isosceles triangles, the angles at the base are equal to one another, and, if the equal sides are extended then the angles under the base are equal to one another.

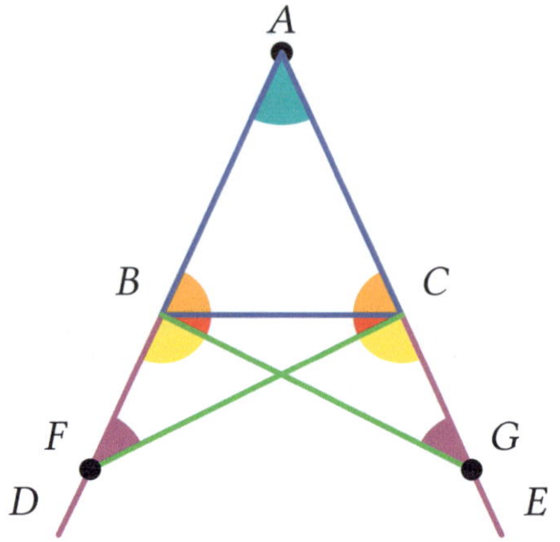

Let ***ABC*** (▲) be an isosceles triangle having the side ***AB*** (——) equal to the side ***AC*** (——), and let the straight-lines ***AB*** (——) and ***AC*** (——) be extended as straight-lines ***BD*** (——) and ***CE*** (——) ***[Post. 2]***.

I say that the angle ***ABC*** (∡) is equal to ***ACB*** (∡), and angle ***CBD*** (∡+∡) is equal to ***BCE*** (∡+∡).

Let a point ***F*** be taken at random on ***BD*** (——), and let ***AG*** (——) be cut off from the greater ***AE*** (——), equal to the lesser ***AF*** (——) ***[Prop. 3]***.

> **Note:** In other words pick a random point ***F*** on ***BD*** (——) and make sure the second point ***G*** on the other leg is at the same relative position on ***CE*** (——).
>
> This is where Proposition 3 becomes useful, it allows Euclid to create lines that are the same length as another line. In this case it's used to make sure that ***AF*** (——) is the same length ***AG*** (——).

Let the straight-lines ***FC*** (——) and ***GB*** (——) be now joined ***[Post. 1]***.

Since ***AF*** (——) is equal to ***AG*** (——), and ***AB*** (——) is equal to ***AC*** (——) , the two straight-lines ***FA*** (——), ***AC*** (——) are equal to the two straight-lines ***GA*** (——), ***AB*** (——), respectively. They also encompass a common angle, ***FAG*** (∡).

> **Note:** The intent here is to create two triangles ***AFC*** and ***AGB***

Thus, the base ***FC*** (——) is equal to the base ***GB*** (——), and the triangle ***AFC*** will be equal to the triangle ***AGB***, and the remaining angles subtendend by the equal sides will be equal to the corresponding remaining angles *[Prop. 4]*.

That is the angle ***ACF*** (∠ + ∠) is the same as ***ABG*** (∠ + ∠), and the angle ***AFC*** (∠) is the same as ***AGB*** (∠).

Since the whole of ***AF*** (————) is equal to the whole of ***AG*** (————), within which ***AB*** (——) is equal to ***AC*** (——), the remainder ***BF*** (——) is thus equal to the remainder ***CG*** (——) [Axiom 3].

But ***FC*** (——) was also shown to be equal to ***GB*** (——). So the two straight-lines ***BF*** (——), ***FC*** (——) are equal to the two straight-lines ***CG*** (——), ***GB*** (——), respectively, and the angle ***BFC*** (∠) is equal to the angle ***CGB*** (∠), and the base ***BC*** (——) is common to them.

> **Note:** We've created another two triangles: ***BFC*** and ***CGB***

Thus, the triangle ***BFC*** will be equal to the triangle ***CGB***, and the remaining angles subtended by the equal sides will be equal to the corresponding remaining angles *[Prop. 4]*.

Thus, ***FBC*** (∠+∠) is equal to ***GCB*** (∠+∠), and ***BCF*** (∠) is equal to ***CBG*** (∠). Therefore, since the whole angle ***ABG*** (∠+∠) was shown to be equal to the whole angle ***ACF*** (∠+∠), within which ***CBG*** (∠) is equal to ***BCF*** (∠), the remainder ***ABC*** (∠) is thus equal to the remainder ***ACB*** (∠) [Axiom 3].

And they are at the base of triangle ***ABC***. And ***FBC*** was also shown to be equal to ***GCB***. And they are under the base.

Thus, for isosceles triangles, the angles at the base are equal to one another, and if the equal sides are produced then the angles under the base will be equal to one another.

Q.E.D.

> **Note:** This is quite a complex proof in terms of the number of angles and edges and will likely take a number of readings to understand it. It involves two pairs of triangles, two applications of SAS and a subtraction through Axiom 3. A useful aid for the reader is to watch one or more of the YouTube videos that describe this proposition.
>
> As a historical note, this proof was called the Asses' Bridge (or *pons asinorum*) and Roger Bacon in 1250 AD also referred to it as Elefuga, or "escape from misery". Apparently Proposition 5 was considered a test, where if one failed to understand Proposition 5, one could avoid the rest of Euclid and thus escape the hard work that was ahead.
>
> Proclus in his commentary noted that Pappus (290-350 AD) provided an alternative and much simpler proof that required no construction. However this alternative only proves $\alpha = \beta$ and not the equality of the angles under the base (△ + △). Appendix 8.2 has a much fuller discussion.

Out of interest I recreated Oliver Byrne's equivalent proof of Proposition 5 below.

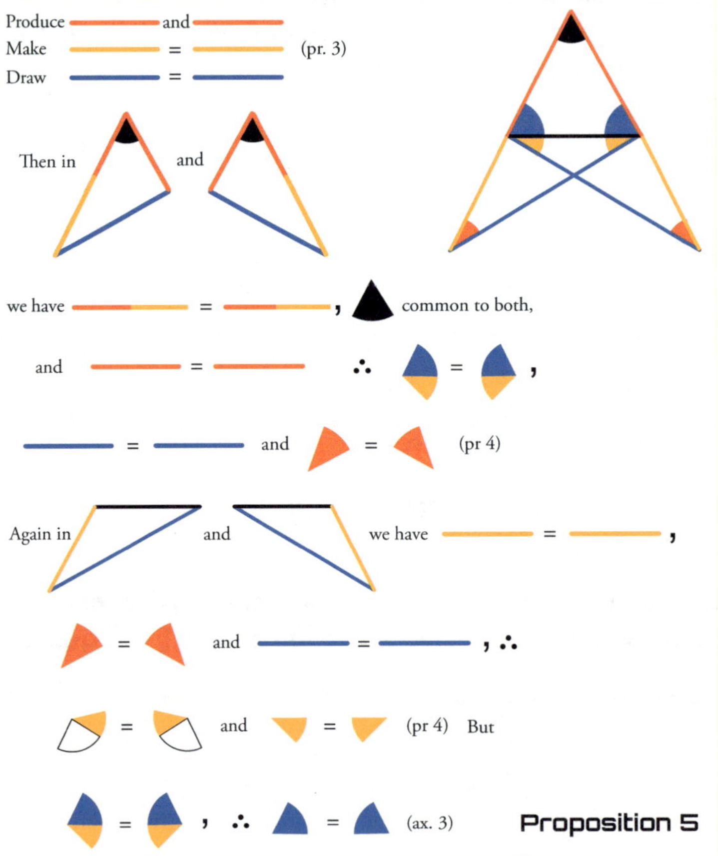

Proposition 6

If a triangle has two angles equal to one another then the sides subtending the equal angles will also be equal to one another.

> **Note:** This proposition is essentially asking the opposite of the last one. It is also the first proposition to use contradiction as a mechanism for proof. This is often referred to as *reductio ad absurdum*.
>
> The proof works by first assuming the lengths are unequal and ends up with a contradiction, suggesting that the assumption that they were unequal is not true, i.e they must be equal.

Let ***ABC*** be a triangle having the angle ***ABC*** (∠) equal to the angle ***ACB*** (∠). I say that side ***AB*** (——) is also equal to side ***AC*** (——).

For if ***AB*** (——) is unequal to ***AC*** (——) then one of them is greater. Let ***AB*** (——) be greater.

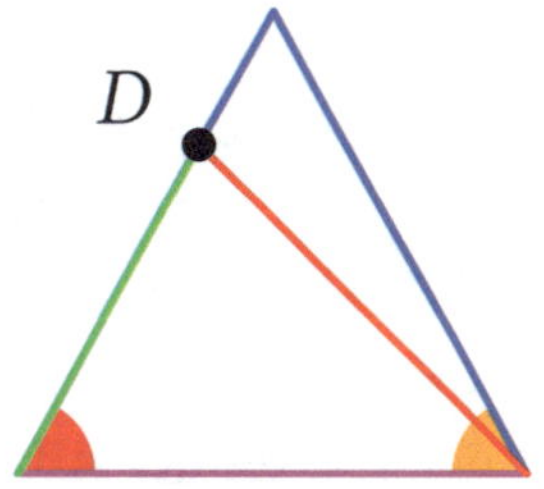

Let ***DB*** (——), equal to the lesser ***AC*** (——), have been cut off from the greater ***AB*** (——) *[Prop. 3]*.

Join the points ***DC*** (——) with a straight-line *[Prop. 1]*.

Therefore, since ***DB*** (——) is equal to ***AC*** (——), and ***BC*** (——) is common, the two sides ***DB*** (——), ***BC*** (——) are equal to the two sides ***AC*** (——), ***CB*** (——), respectively, and the angle ***DBC*** (∠) is equal to the angle ***ACB*** (∠). Thus, the base ***DC*** (——) is equal to the base ***AB*** (——), and the triangle ***DBC*** will be equal to the triangle ***ACB*** *[Prop. 4]*.

The lesser to the greater. The very notion is absurd **[Axiom 5]**.

Thus, ***AB*** (——) is not unequal to ***AC*** (——). Thus, it is equal.

Thus, if a triangle has two angles equal to one another then the sides subtending the equal angles will also be equal to one another.

Q.E.D

> **Note:** One oddity about Proposition 6 is that it is not used at all in Book I. Instead it is required for later books such as II, III, IV, VI and XIII. Maybe it was added to Book I as an introduction to proof by contradiction or was added just to give an example of a complimentary theorem to Proposition 5. Heath, 1925 has a fairly long discussion of this proposition.

Alternative Reading

I find some of the language in Proposition 6 to be unclear, at least on first reading. I therefore took the liberty of creating an alterative rendering of this proof.

Given the triangle ***ABC***, assume that one of the sides is longer than the other, for example:

AB > ***AC***

We will to show that if the triangle is an isosceles triangle, this assertion must be false. Note, although in the figure below it looks like ***AB*** is the same as ***AC***, assume for the moment that visually they are not.

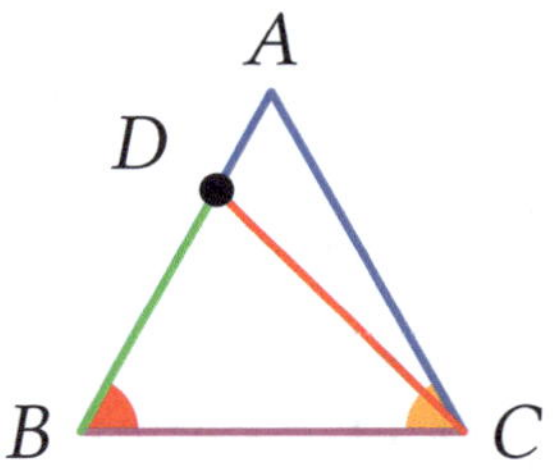

If ***AB*** > ***AC***, lets reduce ***AB*** by an amount so that is matches ***AC***. Let the point where this occurs be labelled point ***D***.

We can now therefore say that ***DB*** = ***AC***

If you look carefully you will see two triangles in the figure above, ***DBC*** and ***ABC***, let's separate them as shown below:

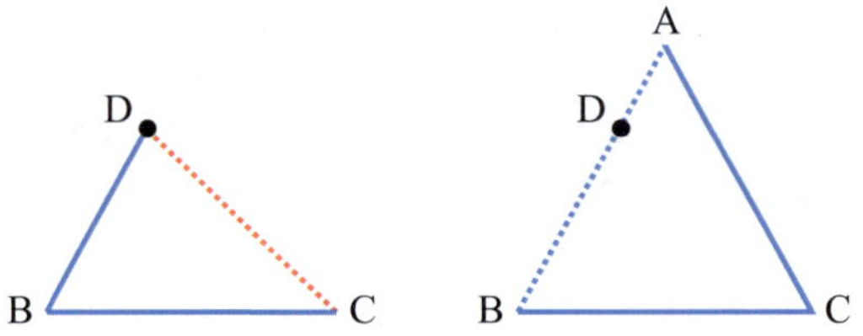

The two triangles have a common base ***BC***. We also stated above that ***DB*** = ***AC***. That is the left two most sides, ***DBC***, and the right two most sides, ***ACB***, are the same, hence the angle between them is the same, ie the angle ***DBC*** () equals the angle ***ACB*** ().

Therefore it must be the case, by SAS (Proposition 4), that the two opposite sides, ***DC*** and łAB must also be the same, i.e

DC = ***AB***. Hence the two triangles, must be identical, ***DBC*** = ***ACB***

But this is nonsense since we had assumed initially that the two triangles were **not** the same. We therefore have a contradiction and the original assertion that ***AB*** > ***AC*** must be false, i.e ***AB*** = ***AC***.

Proposition 7

Given two straight lines constructed from the ends of a straight line and meeting in a point, there cannot be constructed from the ends of the same straight line, and on the same side of it, two other straight lines meeting in another point and equal to the former two respectively, namely each equal to that from the same end.

Note: The objective is to show that the point ***C*** in the figure is unique, given fixed lengths for ***AC*** (—) and ***BC*** (—). It will prove that a point, such as ***D*** in the figure, is impossible if we use the same segments lengths as ***AC*** (—) and ***BC*** (—). It will prove this by supposing we can have a point such as ***D***, which will lead to a contradiction, *reductio ad absurdum*.

If possible, given two straight-lines ***AC*** (—), ***CB*** (—), equal to two other straight-lines ***AD*** (—), ***DB*** (—), respectively, have been constructed on the same straight-line ***AB*** (—), meeting at different points, ***C*** and ***D***, on the same side of ***AB*** (—), and having the same ends on ***AB*** (—).

CA (—) is equal to ***DA*** (—), having the same end ***A***, and ***CB*** (—) is equal to ***DB*** (—), having the same end ***B***.

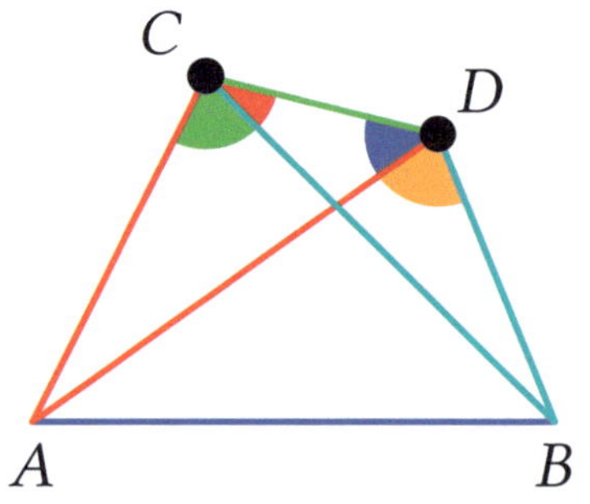

Let ***CD*** (—) have been joined ***[Prop. 1]***.

Therefore, since ***AC*** (—) is equal to ***AD*** (—), the angle ***ACD*** (∠ + ∠) is also equal to angle ***ADC*** (∠) ***[Prop. 5]***.

Thus, ***ADC*** (∠) is greater than ***DCB*** (∠) **[Axiom 5]**.

Thus, ***CDB*** (∠ + ∠) is much greater than ***DCB*** (∠) **[Axiom 5]**.

Since ***CB*** (—) is equal to ***DB*** (—), the angle ***CDB*** (∠ + ∠) is also equal to angle ***DCB*** (∠) ***[Prop. 5]***.

But it was shown that the former angle is also much greater than the latter. The very thing is impossible.

Thus, on the same straight-line, two other straight-lines equal, respectively, to two given straight-lines which meet cannot be constructed meeting at a different point on the same side of the straight-line, but having the same ends as the given straight-lines.

Q.E.D

> **Note:** Euclid doesn't consider the case where point ***D*** is inside the triangle ***ABC***. This is typical of Euclid in that not all cases are considered. This is also the situation with Proposition 2 where only one case is considered but a number of alternatives are possible. As Heath 1925 pointed out, 'This proposition gives Proclus an opportunity, such as the Greek commentators revelled in, of distinguishing a multitude of cases'. Previous editions, particularly post-medieval editions of Euclid tended to fill in the missing cases which made these editions much longer. Why Euclid limited his discussion to just one case is unclear, perhaps it saved on papyrus, perhaps alternative cases were reserved as exercises for the students, which is often the suggested reason.
>
> Proposition 7 is only used once in Book I and that is in the next Proposition 8 which is to prove the SSS theorem.

Proposition 8

If two triangles have two sides equal to two sides respectively, and also have the base equal to the base, then they will also have equal the angles which are contained by the equal straight-lines.

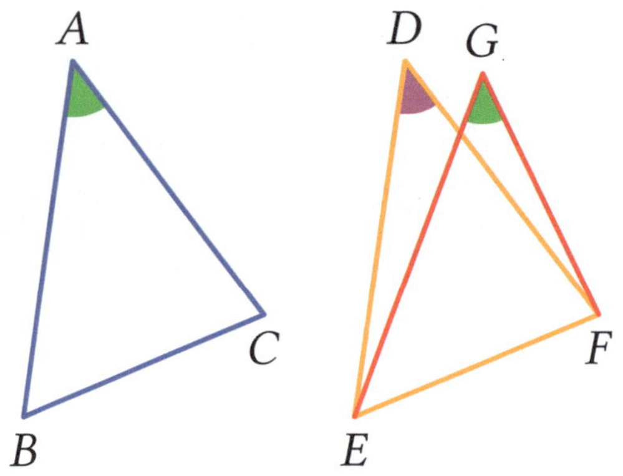

Let ***ABC*** (▲) and ***DEF*** (▲) be two triangles having the two sides ***AB*** (—) and ***AC*** (—) equal to the two sides ***DE*** (—) and ***DF*** (—), respectively. That is ***AB*** (—) to ***DE*** (—), and ***AC*** (—) to ***DF*** (—). Let them also have the base ***BC*** (—) equal to the base ***EF*** (—). I say that the angle ***BAC*** (∠) is also equal to the angle ***EDF*** (∠).

For if triangle ***ABC*** (▲) is applied to triangle ***DEF*** (▲) , the point ***B*** being placed on point ***E***, and the straight-line ***BC*** (—) on ***EF*** (—), then point ***C*** will also coincide with ***F***, on account of ***BC*** (—) being equal to ***EF*** (—) .

Because of ***BC*** (—) coinciding with ***EF*** (—) , the sides ***BA*** (—) and ***CA*** (—) will also coincide with ***ED*** (—) and ***DF*** (—) respectively.

If base ***BC*** (—) coincides with base ***EF*** (—), but the sides ***AB*** (—) and ***AC*** (—) do not coincide with ***ED*** (—) and ***DF*** (—) respectively, but miss like ***EG*** (—) and ***GF*** (—), then we will have constructed upon the same straight-line, two other straight-lines equal, respectively, to two given straight-lines, and meeting at a different point on the same side of the straight-line, but having the same ends. But such straight-lines cannot be constructed ***[Prop. 7]***.

Thus, the base ***BC*** (—) being applied to the base ***EF*** (—), the sides ***BA*** (—) and ***AC*** (—) cannot not coincide with ***ED*** (—) and ***DF*** (—) respectively. Thus, they will coincide. So the angle ***BAC*** () will also coincide with angle ***EDF*** (), and will be equal to it **[Axiom 4]**.

Thus, if two triangles have two sides equal to two sides, respectively, and have the base equal to the base, then they will also have equal the angles which are contained by the equal straight-lines.

Q.E.D

> **Note:** This is the second congruence theorem in Euclid which represents the SSS rule. The SAS rule was described in Proposition 4.
>
> Like Proposition 4, this proposition also uses the less than satisfactory method of superposition. Recall that in Proposition 4, it was mentioned that Hilbert felt that Proposition 4 should be an axiom and in his work he used the SAS axiom in a proof of the SSS (side-side-side) statement.

Proposition 9

To cut a given rectilinear angle in half

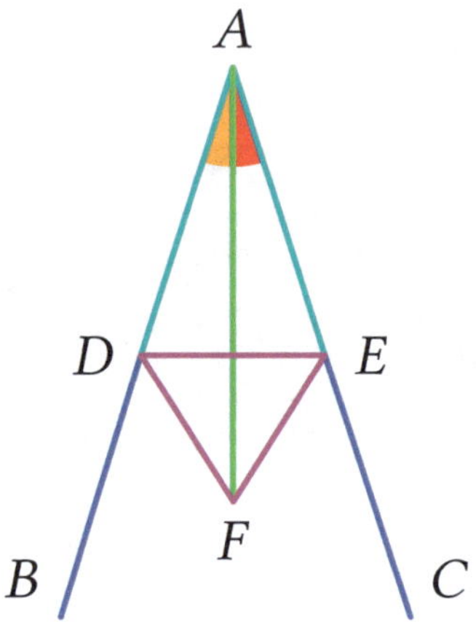

Let ***BAC*** (∠ + ∠) be the given rectilinear angle. So it is required to cut it in half.

Let the point ***D*** be taken at random on ***AB*** (——), and let ***AE*** (——), equal to ***AD*** (——), have been cut off from ***AC*** *[Prop. 3]*

Let ***DE*** (——) be joined, and let the equilateral triangle ***DEF*** (▲) be constructed upon ***DE*** (——) *[Prop. 1]*,

Let ***AF*** (——) be joined. I say that the angle ***BAC*** (∠+∠) has been cut in half by the straight-line ***AF*** (——) .

For since ***AD*** (——) is equal to ***AE*** (——), and ***AF*** (——) is common, the two straight-lines ***DA*** (——), ***AF*** (——) are equal to the two lines ***EA*** (——), ***AF*** (——) , respectively.

The base ***DF*** (——) is equal to the base ***EF*** (——). Thus, angle ***DAF*** (∠) is equal to angle ***EAF*** (∠) *[Prop. 8]*.

Thus, the given rectilinear angle ***BAC*** (∠+∠) has been cut in half by the straight-line ***AF*** (——).

Q.E.F

> **Note:** For the beginner, proposition 9 and 10 are good introductory propositions that are relatively simple to understand although they do require knowledge of the some previous propositions. However the propositions used are not difficult ones.

Proposition 10

To cut a given finite straight-line in half.

Let ***AB*** (——) be the given finite straight-line. It is required to cut the finite straight-line ***AB*** (——) in half.

Let the equilateral triangle ***ABC*** (▲) have been constructed upon ***AB*** (——) *[Prop. 1]*.

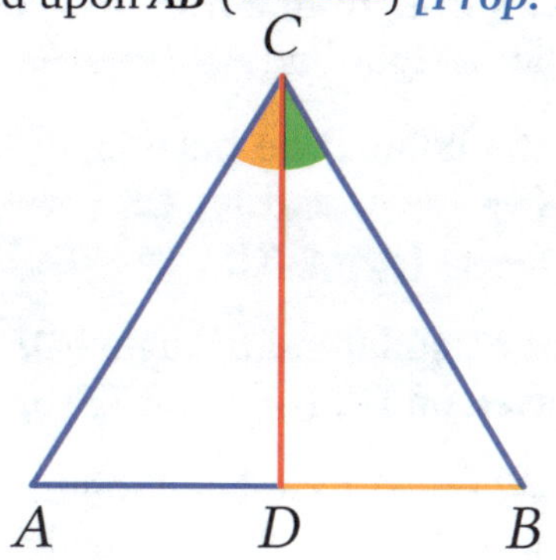

Let the angle ***ACB*** (▲+▲) have been cut in half by the straight-line ***CD*** (——) *[Prop. 9]*.

I say that the straight-line ***AB*** (——) has been cut in half at point ***D***.

Since ***AC*** (——) is equal to ***CB*** (——), and ***CD*** (——) is common, the two straight-lines ***AC*** (——), ***CD*** (——) are equal to the two straight-lines ***BC*** (——) and ***CD*** (——) .

The angle ***ACD*** (▲) is equal to the angle ***BCD*** (▲). Thus, the base ***AD*** (——) is equal to the base ***BD*** (——) *[Prop. 4]*.

Thus, the given finite straight-line ***AB*** (——) has been cut in half at point ***D***.

Q.E.F.

> **Note:** When doing a bisection on paper with a ruler and compass, it's easier to simply draw two circles of radius ***AB***, one centered on ***A*** and the second centered on ***B***. Where the two circles intersect, draw a straight-line between the intersections, this line will intersect line ***AB*** at ***D***.

Proposition 11

To draw a straight-line at right-angles to a given straight-line from a given point on it

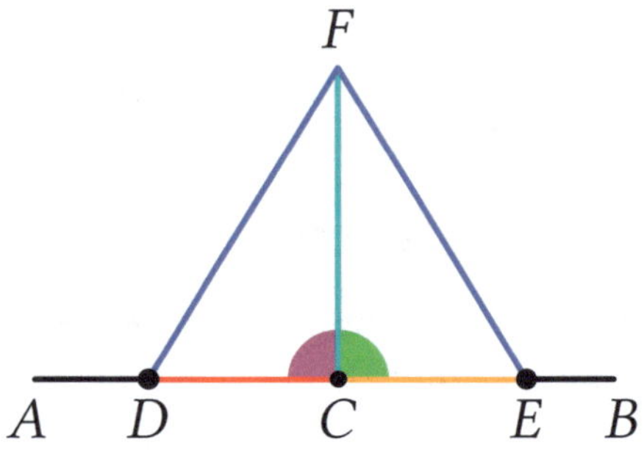

Let ***AB*** (———) be the given straight-line, and ***C*** the given point on it. So it is required to draw a straight-line from the point ***C*** at right-angles to the straight-line ***AB*** (———) .

Let the point ***D*** be have been taken at random on ***AC*** (——), and let ***CE*** (—) be made equal to ***CD*** (—) ***[Prop. 3]***.

Let the equilateral triangle ***FDE*** (▲) have been constructed on ***DE*** (——) ***[Prop. 1]***.

Let ***FC*** (—) have been joined. I say that the straight-line ***FC*** (—) has been drawn at right-angles to the given straight-line ***AB*** (———) from the given point ***C*** on it.

For since ***DC*** (—) is equal to ***CE*** (—), and ***CF*** (—) is common, the two straight-lines ***DC*** (—), ***CF*** (—) are equal to the two straight-lines, ***EC*** (—), ***CF*** (—) , respectively. And the base ***DF*** (—) is equal to the base ***FE*** (—). Thus, the angle ***DCF*** (∠) is equal to the angle ***ECF*** (∠) ***[Prop. 8]***.

That is, they are adjacent. But when a straight-line stood on another straight-line makes the adjacent angles equal to one another, each of the equal angles is a right-angle ***[Def. 10]***.

Thus, each of the angles ***DCF*** (∠) and ***FCE*** (∠) is a right-angle.

Thus, the straight-line ***CF*** (—) has been drawn at right-angles to the given straight-line ***AB*** (———) from the given point ***C*** on it.

Q.E.F.

> **Note:** This construction is very similar to the last one, Proposition 10. The difference is that this time we are using bisection as a means to draw a line at right angles to another line.

Proposition 12

To draw a straight-line perpendicular to a given infinite straight-line from a given point which is not on it.

Note: Notice that the proof for Proposition 11 and 12 is quite similar. In fact, the fifth paragraph in each are identical.

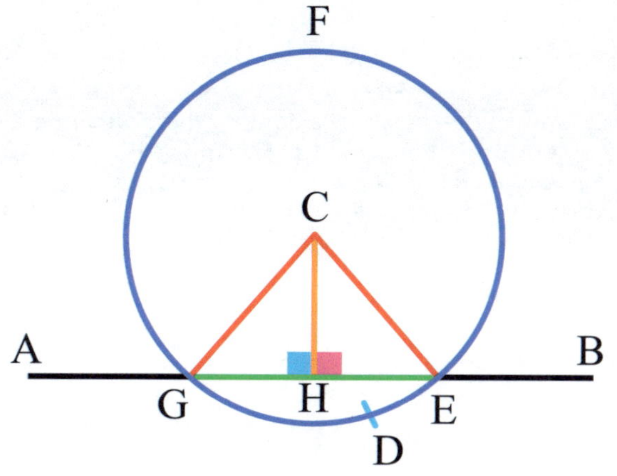

Let ***AB*** (———) be the given infinite straight-line and ***C*** the given point, which is not on ***AB*** (———). So it is required to draw a straight-line perpendicular to the given infinite straight-line ***AB*** (———) from the given point ***C***, which is not on ***AB*** (———).

Let point ***D*** have been taken at random on the other side to ***C*** of the straight-line ***AB*** (———) and let the circle ***EFG*** (●) have been drawn with center ***C*** and radius ***CD*** *[Prop. 3]*.

Let the straight-line ***EG*** have been cut in half at point ***H*** *[Prop. 10]*.

Let the straight-lines ***CG*** (—), ***CH*** (—), and ***CE*** (—) have been joined. I say that the straight-line ***CH*** (—) has been drawn perpendicular to the given infinite straight-line ***AB*** (———) from the given point ***C***, which is not on ***AB*** (———).

For since ***GH*** (—) is equal to ***HE*** (—), and ***HC*** (—) is common, the two straight-lines ***GH*** (—), ***HC*** (—) are equal to the two straight-lines ***EH*** (—), ***HC*** (—), respectively, and the base ***CG*** (—) is equal to the base ***CE*** (—) . Thus, the angle ***CHG*** (∠) is equal to the angle ***EHC*** (∠) ***[Prop. 8]***.

That is, they are adjacent. But when a straight-line stood on another straight-line makes the adjacent angles equal to one another, each of the equal angles is a right-angle, and the former straight-line is called a perpendicular to that upon which it stands **[Axiom 10]**.

Thus, the straight-line ***CH*** (—) has been drawn perpendicular to the given infinite straight-line ***AB*** (—) from the given point ***C***, which is not on ***AB*** (—).

Q.E.F

> **Note:** This is a slight variation on the last Proposition 11 where we instead draw a perpendicular line based on a point, ***C***, that is not on the line ***AB***. This proposition isn't used in Book I but only in later books. It might have been added here to satisfies the Greek astronomers who it is claimed found it useful (Proclus, ~450AD, Heath, 1925).

Proposition 13

If a straight-line stood on another straight-line makes angles, it will certainly either make two right-angles, or angles whose sum is equal to two right-angles.

> **Note:** Although the proof for this proposition is somewhat long, the theorem is saying something that would appear to be self evident. However, it is the first example where Euclid has statements on adding angles. The proposition compares two cases, the first when ***BE*** (——) is perpendicular to ***DC*** (——), hence on either side we have a right angle, and a second case when ***BA*** (——) is leaning, say to the right (see figure), then the sum of the angles, though not right angles anymore, still add up to two right angle's worth.

Let some straight-line ***AB*** (——) be stood on the straight-line ***CD*** (——) that makes the angles ***CBA*** (∠) and ***ABD*** (∠+∠). I say that the angles ***CBA*** (∠) and ***ABD*** (∠+∠) are certainly either two right-angles, or have a sum equal to two right-angles.

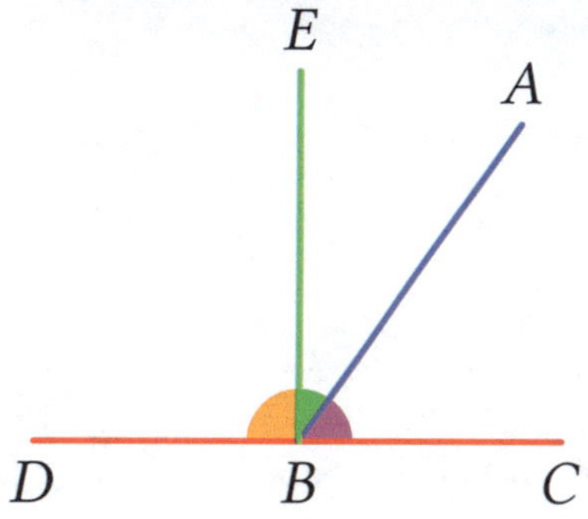

In fact, if ***CBA*** (∠) is equal to ***ABD*** (∠+∠) then they are two right-angles **[Axiom 10]**.

But, if not, let ***BE*** (——) have been drawn from the point ***B*** at right-angles to the straight-line ***CD*** (——) ***[Prop. 11]***.

Thus, ***CBE*** (∠+∠) and ***EBD*** (∠) are two right-angles.

Since ***CBE*** (∠+∠) is equal to the sum of the two angles ***CBA*** (∠) and ***ABE*** (∠), let ***EBD*** (∠) have been added to both. Thus, the sum of the angles ***CBE*** (∠+∠) and ***EBD*** (∠) is equal to the sum of the three angles to ***CBA*** (∠), ***ABE*** (∠), and ***EBD*** (∠) **[Axiom 2]**.

Again, since ***DBA*** (∠+∠) is equal to the sum of the two angles ***DBE*** (∠) and ***EBA*** (∠), let ***ABC*** (∠) have been added to both. Thus, the sum of the angles ***DBA*** (∠+∠) and ***ABC*** (∠) is equal to the sum of the three angles ***DBE*** (∠), ***EBA*** (∠), and ***ABC*** (∠) **[Axiom 2]**.

But the sum of ***CBE*** (∠+∠) and ***EBD*** (∠) was also shown to be equal to the sum of the same three angles. And things equal to the same thing are also equal to one another **[Axiom 1]**.

Therefore, the sum of ***CBE*** (∠+∠) and ***EBD*** (∠) is also equal to the sum of ***DBA*** (∠+∠) and ***ABC*** (∠). But, the sum of ***CBE*** (∠+∠) and ***EBD*** (∠) is two right-angles. Thus, the sum of ***ABD*** (∠+∠) and ***ABC*** (∠) is also equal to two right-angles.

Thus, if a straight-line stood on another straight-line makes angles, it will certainly either make two right-angles, or angles whose sum is equal to two right-angles. Q.E.D

> **Note:** This is quite a convoluted proof for something that is simple. Its ultimate conclusion is that if we have multiple lines standing on the same sides of a horizontal line at a point, ***B***, but leaning in different directions, the sum of all the angles between the lines will be 180°.

> **Note:** Here is a reinterpretation of the previous proposition using algebra. The use of algebra makes the proof much simpler and gives a hint as to the limits of a purely geometric approach.
>
> If we label the three angles in the figure as ***CBA*** (∠) $= \alpha$, ***ABE*** (∠) $= \beta$, and ***EBD*** (∠) $= \gamma$, then what the proposition says is: $\alpha + \beta + \gamma = 180^o$, which I think you might agree is somewhat obvious. Nevertheless, such a statement can be proved as done in this proposition. In more modern notation the proof is a follows.
>
> We know by construction that $\gamma = ∟$ and $\alpha + \beta = ∟$
>
> We also know the following two statements to be true:
>
> (1) $\gamma + (\alpha + \beta) = \gamma + \alpha + \beta$ ∠ + (∠ + ∠) = ∠ + ∠ + ∠
>
> (2) $\alpha + (\gamma + \beta) = \alpha + \gamma + \beta$ ∠ + (∠ + ∠) = ∠ + ∠ + ∠
>
> However, the right-hand sides of (1) and (2) are equal, therefore:
>
> $\gamma + (\alpha + \beta) = \alpha + (\gamma + \beta)$ ∠ + (∠ + ∠) = ∠ + (∠ + ∠)
>
> But from above $\gamma = ∟$ (i.e ∠ = ∟), and $\alpha + \beta = ∟$, (∠ + ∠ = ∟) therefore:
>
> $∟ + ∟ = \alpha + (\gamma + \beta)$ ∟ + ∟ = ∠ + (∠ + ∠)
>
> Which is what we set out to show. Note that Euclid considers 180° to be two right-angles. The Greeks didn't have a name for a 180° angle other than the phrase 'straight-on'. This proposition can be generalized to any number of lines standing on ***CD***.

Proposition 14

If with any straight line, and at a point on it, two straight lines not lying on the same side (he means ***BC*** and ***BD***) make the adjacent angles equal to two right angles, the two straight lines will be in a straight line with one another.

For with any straight line ***AB*** (—), and at the point ***B*** on it, let the two straight lines ***BC*** (—), ***BD*** (—) not lying on the same side make the adjacent angles ***ABC*** (△), ***ABD*** (△ + △) equal to two right angles.

I say that ***BD*** (—) is in a straight line with ***CB*** (—) .

> **Note:** By 'straight line', Euclid means the lines are 180 degree with respect to each other.

For if ***BD*** (—) is not in a straight line with ***BC*** (—) then let ***BE*** (—) be in a straight line with ***CB*** (—).

> **Note:** Euclid is proposing another line ***BE*** (—) that is 180° degrees to line ***BC*** (—). Obviously in the figure it doesn't look like that.

Then, since the straight line ***AB*** (—) stands on the straight line ***CBE*** (——), the angles ***ABC*** (△), ***ABE*** (△) are equal to two right angles ***[Prop. 13]***.

> **Note:** This is what Euclid is asserting to show is false. Obviously in the diagram it doesn't look like ***BE*** (—) is 180° to ***CB*** (—).

But the angles ***ABC*** (△) , ***ABD*** (△ + △) are also equal to two right angles, therefore the angles ***CBA*** (△) and ***ABE*** (△) are equal to the angles ***CBA*** (△) and ***ABD*** (△+△).

Let the angle ***CBA*** (△) be subtracted from each. Therefore the remaining angle ***ABE*** (△) is equal to the remaining angle ***ABD*** (△+△) **[Axiom 3]**.

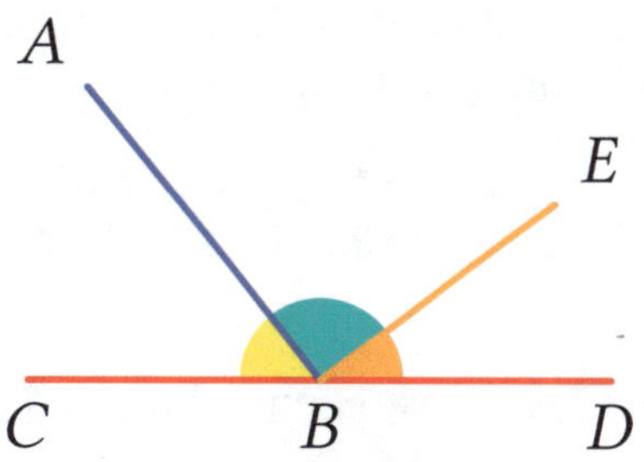

The less to the greater. which is impossible.

Therefore, ***BE*** (—) is not in a straight line with respect to ***CB*** (—). Similarly, we can prove that neither is any other straight-line except ***BD*** (—).

Thus, ***CB*** (—) is a straight line with respect to ***BD*** (—).

Thus, if with any straight line, and at a point on it, two straight lines not lying on the same side make the adjacent angles equal to two right angles, the two straight lines will be in a straight line with one another.

Q.E.D

Proposition 15

If two straight-lines cut one another then they make the vertically opposite angles equal to one another.

> **Note:** The use of the phrase "vertically opposite angles" means angles opposite each other, for example ∠ and ∠ are opposite and so are angles ∠ and ∠. This terminology is still used today in geometry.

Let the two straight-lines ***AB*** (—) and ***CD*** (—) cut one another at the point ***E***. I say that angle ***AEC*** (∠) is equal to angle ***DEB*** (∠), and angle ***CEB*** (∠) to angle ***AED*** (∠).

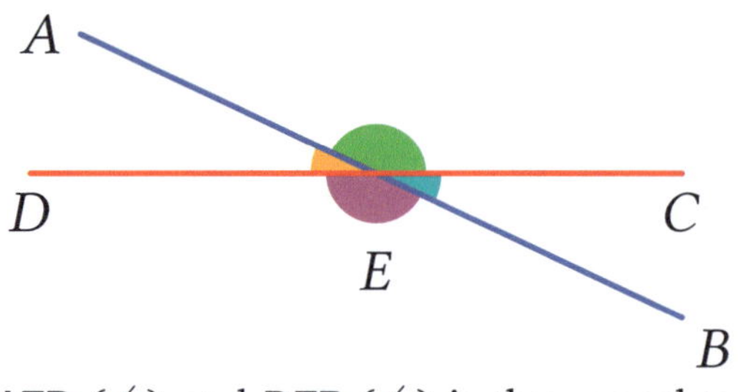

For since the straight-line ***AE*** stands on the straight-line ***CD*** (—), making the angles ***CEA*** (∠) and ***AED*** (∠), the sum of the angles ***CEA*** (∠) and ***AED*** (∠) is thus equal to two right-angles ***[Prop. 13]***.

Again, since the straight-line ***DE*** stands on the straight-line ***AB*** (—), making the angles ***AED*** (∠) and ***DEB*** (∠), the sum of the angles ***AED*** (∠) and ***DEB*** (∠) is thus equal to two right-angles ***[Prop. 13]***.

But the sum of ***CEA*** (∠) and ***AED*** (∠) was also shown to be equal to two right-angles. Thus, the sum of ***CEA*** (∠) and ***AED*** (∠) is equal to the sum of ***AED*** (∠) and ***DEB*** (∠) **[Axiom 1]**.

Let ***AED*** (∠) have been subtracted from both. Thus, the remainder ***CEA*** (∠) is equal to the remainder ***BED*** (∠) **[Axiom 3]**.

Similarly, it can be shown that ***CEB*** () and ***DEA*** () are also equal.

Thus, if two straight-lines cut one another then they make the vertically opposite angles equal to one another.

Q.E.D

Note: This is probably one of the first proofs that a middle or high School student will learn. Interestingly, in Euclid, it's a later proposition due to the way the propositions are set up.

Using algebra, the proof is quite simple. We know that  +  = 180°. We also know that  +  = 180°. Hence  +  =  + . We cancel the blues leaving:  =  which is the result we seek. We can use the same logic to show that  = .

Proposition 16

For any triangle, when one of the sides is produced, the external angle is greater than each of the internal and opposite angles.

> This is another fairly involved proof but well worth trying to understand it. It's a clever use of a variety of previous propositions.

Let ***ABC*** (▲) be a triangle, and let one of its sides ***BC*** (—) have been produced to ***D***. I say that the external angle ***ACD*** (∡+∡) is greater than each of the internal and opposite angles, ***CBA*** (∡ + ∡) and ***BAC*** (∡).

Let the straight-line ***AC*** (—) have been cut in half at point ***E*** ***[Prop. 10]***.

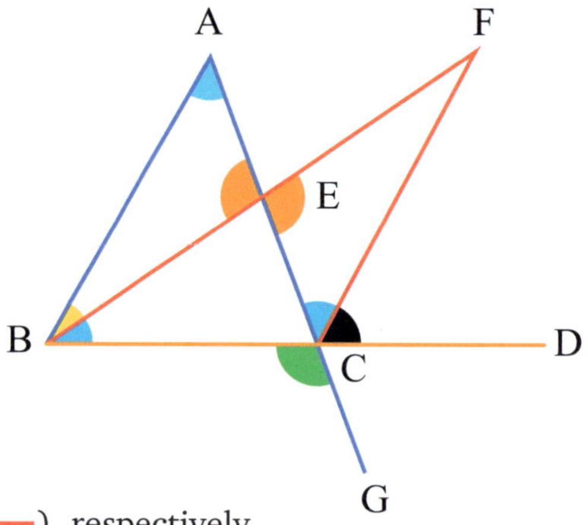

And ***BE*** (—) being joined, let it have been produced in a straight-line to point ***F***.

Let ***EF*** (—) be made equal to ***BE*** (—) ***[Prop. 3]***.

Let ***FC*** (—) have been joined, and let ***AC*** (—) have been drawn through to point ***G***.

Therefore, since ***AE*** (—) is equal to ***EC*** (—), and ***BE*** (—) to ***EF*** (—), the two straight-lines ***AE*** (—), ***EB*** (—) are equal to the two straight-lines ***CE*** (—), ***EF*** (—), respectively.

Also, angle ***AEB*** (∡) is equal to angle ***FEC*** (∡), for they are vertically opposite ***[Prop. 15]***.

Thus, the base ***AB*** (—) is equal to the base ***FC*** (—), and the triangle ***ABE*** is equal to the triangle ***FEC*** (▲), and the remaining angles subtended by the equal sides are equal to the corresponding remaining angles ***[Prop. 4]***.

Thus, ***BAE*** (∡) is equal to ***ECF*** (∡). But ***ECD*** (∡+∡) is greater than ***ECF*** (∡). Thus, ***ACD*** (∡+∡) is greater than ***BAE*** (∡).

Similarly, by having cut ***BC*** (—) in half, it can be shown that ***BCG*** (∡), that is to say,

ACD (∡+∡) is also greater than ***ABC*** (∡+∡).

Thus, for any triangle, when one of the sides is produced, the external angle is greater than each of the internal and opposite angles. Q.E.D

Proposition 17

For any triangle, the sum of two angles taken together in any possible way is less than two right-angles.

Note: Sometimes also known as the exterior angle theorem. This says that the sum any combination of two internal angles in a triangle will always be less than 180°. For example: ∡ + ∡ < 180°. This is a natural follow on from Proposition 16, and is a nice result, but surprisingly isn't used anywhere else in Book I.

Let ***ABC*** (▲) be a triangle. I say that the sum of two angles of triangle ***ABC*** (▲) taken together in any possible way is less than two right-angles.

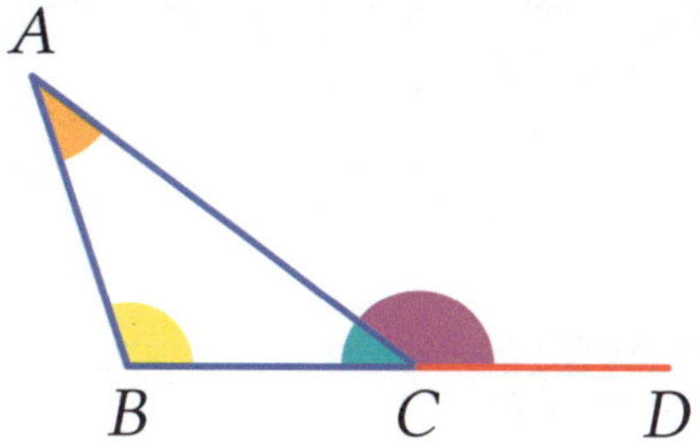

For let ***BC*** (—) have been produced to ***D***.

And since the angle ***ACD*** (∡) is external to triangle ***ABC*** (▲) , it is greater than the internal and opposite angle ***ABC*** (∡) *[Prop. 16]*.

Let ***ACB*** (∡) have been added to both. Thus, the sum of the angles ***ACD*** (∡) and ***ACB*** (∡) is greater than the sum of the angles ***ABC*** (∡) and ***BCA*** (∡). But, the sum of ***ACD*** (∡) and ***ACB*** (∡) is equal to two right-angles *[Prop. 13]*.

Thus, the sum of ***ABC*** (∡) and ***BCA*** (∡) is less than two right-angles. Similarly, we can show that the sum of ***BAC*** (∡) and ***ACB*** (∡) is also less than two right-angles, and further that the sum of ***CAB*** (∡) and ***ABC*** (∡) is less than two right-angles.

Thus, for any triangle, the sum of two angles taken together in any possible way is less than two right-angles. Q.E.D

Note: This is quite a straight forward proof if we use modern notation. First assign the following symbols to the angles: ***ACD*** (▲) = σ, ***ACB*** (▲) = γ, and ***ABC*** (▲) = β.

We first note that $\sigma > \beta$ (Prop. 16), we add γ to both sides to give:

$\sigma + \gamma > \beta + \gamma$

But $\sigma + \gamma = 180^o$, therefore:

$180^o > \beta + \gamma$. That is the sum of the two internal angles is less than 180°.

Proposition 18

In any triangle, the greater side subtends the greater angle.

Note: The title is quite terse, so let's expand it to: "The angle opposite the longer side of a triangle will be the largest angle." The proof only considers one of the angles, in this case ***BCA*** (▲) but the mirror image of the proof also works for the other angle ***BAD***. Hence the angle opposite the longer side will be the largest angle of the three.

You can seen from the current crop of Propositions that Euclid is working through basic properties of triangles.

For let ***ABC*** (▲) be a triangle having side ***AC*** (——) greater than ***AB*** (——). I say that angle ***ABC*** (▲ + ▲) is also greater than ***BCA*** (▲).

For since ***AC*** (——) is greater than ***AB*** (——), let ***AD*** (——) be made equal to ***AB*** (——) ***[Prop. 3]***.

Let ***BD*** (——) have been joined.

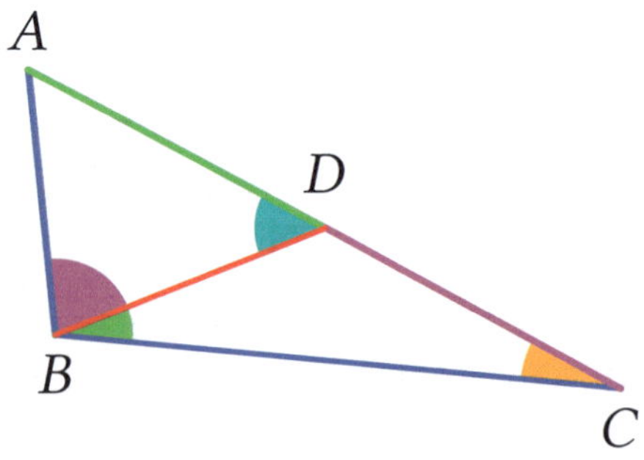

Since angle ***ADB*** (▲) is external to triangle ***BCD***, it is greater than the internal and opposite angle ***DCB*** (▲) ***[Prop. 16]***.

But ***ADB*** (▲) is equal to ***ABD*** (▲), since side ***AB*** (——) is also equal to side ***AD*** (——) ***[Prop. 5]***.

Thus, ***ABD*** (▲) is also greater than ***ACB*** (▲). Thus, ***ABC*** (▲+▲) is much greater than ***ACB*** (▲).

Thus, in any triangle, the greater side subtends the greater angle. Q.E.D

Proposition 19

In any triangle, the greater angle is subtended by the greater side.

Let ***ABC*** be a triangle having the angle ***ABC*** (∠) greater than ***BCA*** (∠). I say that side ***AC*** is also greater than side ***AB***.

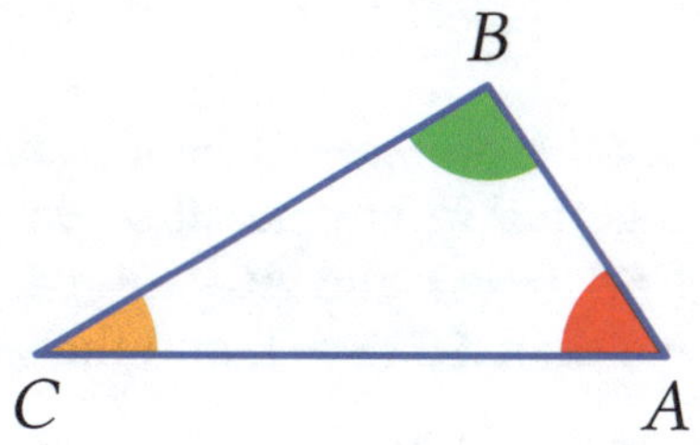

For if not, ***AC*** is certainly either equal to, or less than, ***AB***. In fact, ***AC*** is not equal to ***AB***. For then angle ***ABC*** (∠) would also have been equal to ***ACB*** (∠) ***[Prop. 5]***.

But it is not. Thus, ***AC*** is not equal to ***AB***. Neither, indeed, is ***AC*** less than ***AB***. For then angle ***ABC*** (∠) would also have been less than ***ACB*** (∠) ***[Prop. 18]***.

But it is not. Thus, ***AC*** is not less than ***AB***. But it was shown that ***AC*** is not equal to ***AB*** either. Thus, ***AC*** is greater than ***AB***.

Thus, in any triangle, the greater angle is subtended by the greater side.

Q.E.D

Note: This proposition is the converse of the last one (18); that is, if we pick the largest angle in a triangle, the side opposite this angle will be the longest. Although both (18) and (19) are dealing with related questions, (18) gives a direct proof whereas (19) uses contradiction as the mechanism of proof. Unlike the usual *reductio ad absurdum*, this posits three alternatives of which only one can be true. It successively disproves two, leaving one left which must therefore be the truth. Later commentators such as Proclus gave a direct proof of this proposition but it requires another related result which Euclid didn't have at this stage. Personally I prefer direct proofs as I think they tend to give more insight into the problem.

Proposition 20

In any triangle, the sum of two sides taken together in any possible way is greater than the remaining side.

> **Note:** This is an interesting proposition because it basically proves (at least in Euclidian space) that the shortest route to a point is a straight line. As you can see it takes quite a bit of effort to prove what seems an obvious assertion.

For let ***ABC*** (———) be a triangle. I say that in triangle ***ABC*** the sum of two sides taken together in any possible way is greater than the remaining side. Therefore, the sum of ***BA*** (—) and ***AC*** (—) is greater than ***BC*** (—), the sum of ***AB*** (—) and ***BC*** (—) than ***AC*** (—), and the sum of ***BC*** (—) and ***CA*** (—) than ***AB*** (—).

For let ***BA*** (—) have been drawn through to point ***D***, and let ***AD*** (—) be made equal to ***CA*** (—) *[Prop. 3]*.

Let ***DC*** (—) have been joined.

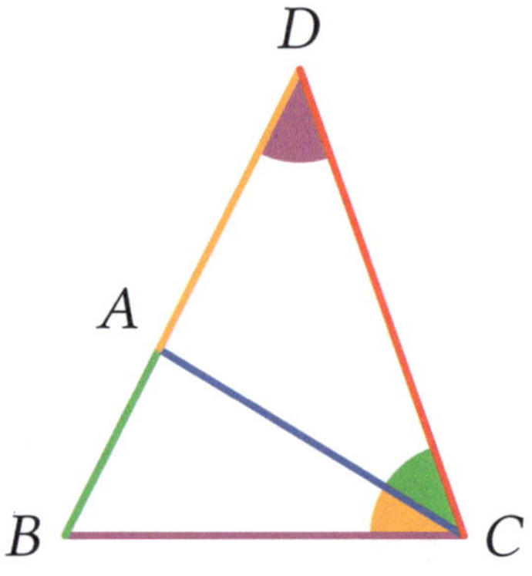

Therefore, since ***DA*** (—) is equal to ***AC*** (—), the angle ***ADC*** (∡) is also equal to ***ACD*** (∡) *[Prop. 5]*.

Thus, ***BCD*** (∡+∡) is greater than ***ADC*** (∡). Since ***DCB*** (∡+∡) is a triangle having the angle ***BCD*** (∡+∡) greater than ***BDC*** (∡), and the greater angle subtends the greater side *[Prop. 19]*.

DB (——) is thus greater than ***BC*** (—). But ***DA*** (—) is equal to ***AC*** (—). Thus, the sum of ***BA*** (—) and ***AC*** (—) is greater than ***BC*** (—). Similarly, we can show that the sum of ***AB*** (—) and ***BC*** (—) is also greater than ***CA*** (—), and the sum of ***BC*** (—) and ***CA*** (—) is greater than ***AB*** (—).

Thus, in any triangle, the sum of two sides taken together in any combination is greater than the remaining side. Q.E.D

Proposition 21

If two internal straight-lines are constructed on one of the sides of a triangle, from its ends, the constructed straight-lines will be less than the two remaining sides of the triangle, but will encompass a greater angle.

Note: This proposition is about the properties of a smaller triangle, such as ▲, that is embedded inside larger triangle, ▲. The Proposition proves that the angle at the tip (∠) of the larger triangle will always be smaller than the angle at the tip (∠) of the smaller triangle. In addition, the sum of the two vertical sides of the outer triangle will always be longer then the sum of the vertical sides of the inner triangle. If you think about it, both results should seem obvious and Proposition 21 proves it.

For let the two internal straight-lines ***BD*** (——) and ***DC*** (——) have been constructed on one of the sides ***BC*** (——) of the triangle ***ABC*** (▲), from its ends ***B*** and ***C*** respectively. I say that ***BD*** (——) and ***DC*** (——) are less than the sum of the two remaining sides of the triangle ***BA*** (——) and ***AC*** (——), but encompass an angle ***BDC*** (∠) greater than ***BAC*** (∠).

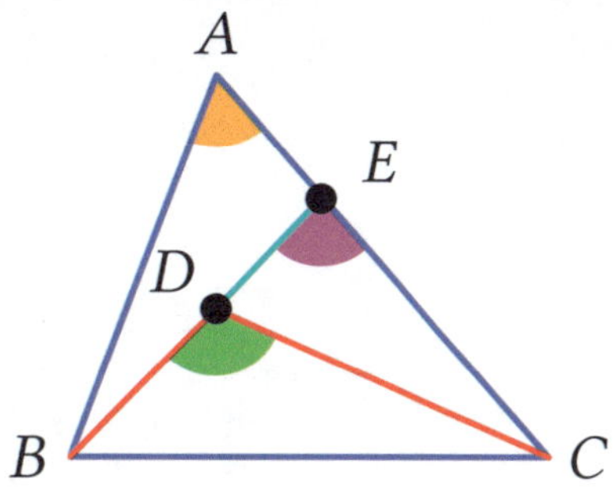

Let ***BD*** (——) have been drawn through to ***E***.

Since in any triangle the sum of any two sides is greater than the remaining side *[Prop. 20]*.

In triangle ***ABE*** the sum of the two sides ***AB*** (——) and ***AE*** (——) is thus greater than ***BE*** (————). Let ***EC*** (——) have been added to both. Thus, the sum of ***BA*** (——) and ***AC*** (——) is greater than the sum of ***BE*** (————) and ***EC*** (——). Again, since in triangle ***CED*** the sum of the two sides ***CE*** (——) and ***ED*** (——) is greater than ***CD*** (——), let ***DB*** (——) have been added to both.

Thus, the sum of ***CE*** (——) and ***EB*** (————) is greater than the sum of ***CD*** (——) and ***DB*** (——). But, the sum of ***BA*** (——) and ***AC*** (——) was shown to be greater than the sum of ***BE*** (————) and ***EC*** (——). Thus, the sum of ***BA*** (——) and ***AC*** (——) is much greater than the sum of ***BD*** (——) and ***DC*** (——).

Again, since in any triangle the external angle is greater than the internal and opposite angles *[Prop. 16]*.

In triangle ***CDE*** the external angle ***BDC*** (∠) is thus greater than ***CED*** (∠). Accordingly, for the same reason, the external angle ***CEB*** (∠) of the triangle ***ABE*** is also greater than ***BAC*** (∠). But, ***BDC*** (∠) was shown to be greater than ***CEB*** (∠). Thus, ***BDC*** (∠) is much greater than ***BAC*** (∠).

Thus, if two internal straight-lines are constructed on one of the sides of a triangle, from its ends, the constructed straight-lines are less than the two remaining sides of the triangle, but encompass a greater angle.

Q.E.D

Proposition 22

To construct a triangle from three straight-lines which are equal to three given straight-lines. It is necessary for the sum of two of the straight-lines taken together in any possible way to be greater than the remaining one.

Note: This is a proposition which describes how to build a triangle starting with three straight lines of different lengths. Proposition 20 puts a restriction on the three straight lines we can use to build a triangle and showed that the sum of any two sides of a triangle will be greater than the length of the remaining side. As a result we must ensure that the three straight lines follow this rule in order for Proposition 22 to work.

Proposition 22 is elegant in its simplicity. We line up the three straight lines one after the other, and we draw two circles centered on the first and second joins in the lines, with radii given by the first line and third line. The triangle falls out naturally from the intersection of the two circles. The proof is also quite simple.

Let ***A***, ***B***, and ***C*** be the three given straight-lines (——), of which let the sum of two taken together in any combination be greater than the remaining line. That is:

the sum of ***A*** and ***B*** is greater than ***C***,
the sum of ***A*** and ***C*** than ***B***,
and the sum of ***B*** and ***C*** than ***A***.

It is required to construct a triangle from straight-lines equal to ***A***, ***B***, and ***C***.

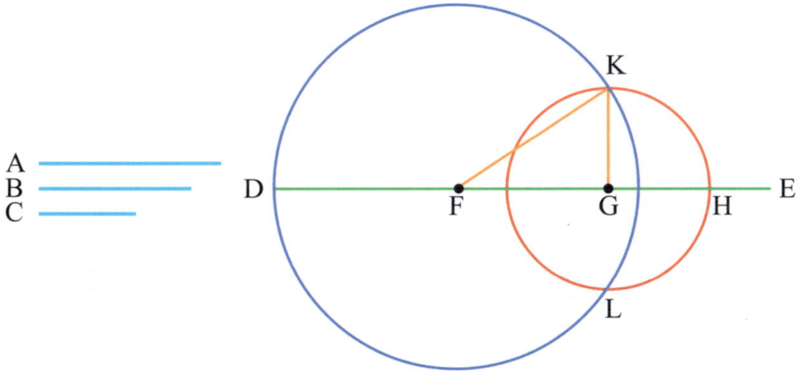

Let some straight-line ***DE*** (——) be set out, terminated at ***D***, and infinite in the direction of ***E***.

Let ***DF*** be made equal to ***A***, and ***FG*** equal to ***B***, and ***GH*** equal to ***C*** *[Prop. 3]*.

Draw a circle ***DKL*** (●) with center ***F*** and radius ***FD***.

Draw a circle ***KLH*** (●) with center ***G*** and radius ***GH***.

Let ***KF*** and ***KG*** be joined. I say that the triangle ***KFG*** (▲) has been constructed from three straight-lines equal to ***A***, ***B***, and ***C***.

Since point ***F*** is the center of the circle ***DKL*** (●), ***FD*** is equal to ***FK***. But, ***FD*** is equal to ***A***. Thus, ***KF*** is also equal to ***A***.

Since point ***G*** is the center of the circle ***LKH*** (●), ***GH*** is equal to ***GK***. But, ***GH*** is equal to ***C***. Thus, ***KG*** is also equal to ***C***. And ***FG*** is also equal to ***B***. Thus, the three straight-lines ***KF***, ***FG***, and ***GK*** are equal to ***A***, ***B***, and ***C*** respectively.

Thus, the triangle ***KFG*** (▲) has been constructed from the three straight-lines ***KF***, ***FG***, and ***GK***, which are equal to the three given straight-lines ***A***, ***B***, and ***C*** respectively.

Q.E.F

Proposition 23

To construct a rectilinear angle equal to a given rectilinear angle at a given point on a given straight-line.

Note: This proposition describes a means to duplicate an angle at another location. At first glance one might think this would be quite difficult but it turns out to be straight forward. It relies on the previous proposition that takes a set of three lines and constructs a triangle. This uses the same approach but the three lines this time are the lines that come from the triangle we want to copy. The angle we to copy is the green angle (∠).

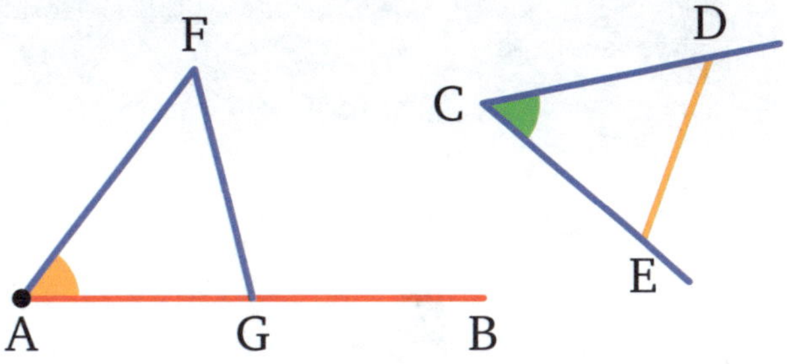

Let ***AB*** (—) be the given straight-line, and ***A*** the given point on it, and ***DCE*** (∠) the given rectilinear angle. So it is required to construct a rectilinear angle equal to the given rectilinear angle ***DCE*** (∠) at the given point ***A*** on the given straight-line ***AB*** (—).

Let the points ***D*** and ***E*** have been taken at random on each of the straight-lines ***CD*** and ***CE*** respectively, and let ***DE*** (—) have been joined. And let the triangle ***AFG*** have been constructed from three straight-lines which are equal to ***CD***, ***DE*** (—), and ***CE***, such that ***CD*** is equal to ***AF***, ***CE*** to ***AG***, and further ***DE*** (—) to ***FG*** *[Prop. 22]*.

Therefore, since the two straight-lines ***DC***, ***CE*** are equal to the two straight-lines ***FA***, ***AG***, respectively, and the base ***DE*** (—) is equal to the base ***FG***, the angle ***DCE*** (∠) is thus equal to the angle ***FAG*** (∠) *[Prop. 8]*.

Thus, the rectilinear angle ***FAG*** (∠), equal to the given rectilinear angle ***DCE*** (∠), has been constructed at the given point ***A*** on the given straight-line ***AB*** (—) .

Q.E.F

Proposition 24

If two triangles have two sides equal to two sides, respectively, but one has the angle encompassed by the equal straight-lines greater than the corresponding angle in the other, then the former triangle will also have a base greater than the base of the latter

Note: This proves something that would appear obvious by inspection, but it takes a surprising amount of effort to prove. Take two triangles, positioned in a column, one above the other. The upper is splayed out wider than the other, i.e the angle at the tip is wider. Hence, given equal vertical sides, the base side on the splayed out triangle will be wider than the less splayed out triangle. Note the relationship with Proposition 21 which is about the relative size of two angles and sides in two triangle. In contrast, this proposition looks instead at the length of the base in a related situation.

This proposition requires the previous proposition that copied an angle.

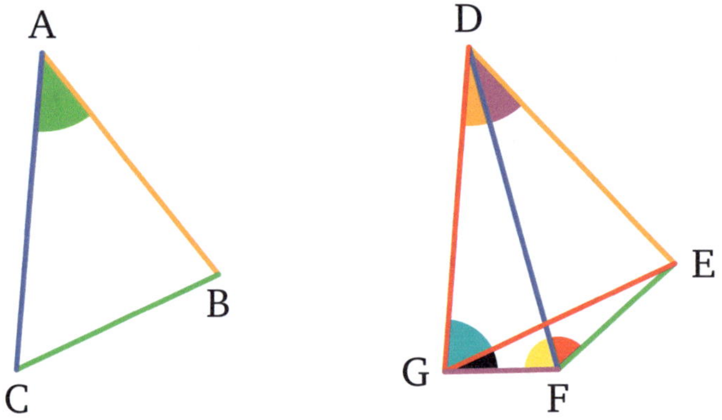

Let ***ABC*** and ***DEF*** be two triangles having the two sides ***AB*** (——) and ***AC*** (——) equal to the two sides ***DE*** (——) and ***DF*** (——), respectively. That is, ***AB*** (——) equal to ***DE*** (——), and ***AC*** (——) to ***DF*** (——). Let them also have the angle at ***A*** (∠) greater than the angle at ***D*** (∠+∠).

I say that the base ***BC*** (——) is also greater than the base ***EF*** (——).

For since angle ***BAC*** (∠) is greater than angle ***EDF*** (∠), let angle ***EDG*** (∠+∠), equal to angle ***BAC*** (∠), have been constructed at the point ***D*** on the straight-line ***DE*** (——) *[Prop. 23]*.

Let ***DG*** (—) be made equal to either of ***AC*** (—) or ***DF*** (—) *[Prop. 3]*,

Let ***EG*** (—) and ***FG*** (—) have been joined.

Therefore, since ***AB*** (—) is equal to ***DE*** (—) and ***AC*** (—) to ***DG*** (—), the two straight-lines ***BA*** (—), ***AC*** (—) are equal to the two straight-lines ***ED*** (—), ***DG*** (—), respectively.

Also the angle ***BAC*** (▲) is equal to the angle ***EDG*** (▲+▲). Thus, the base ***BC*** (—) is equal to the base ***EG*** (—) *[Prop. 4]*.

Again, since ***DF*** (—) is equal to ***DG*** (—), angle ***DGF*** (▲+▲) is also equal to angle ***DFG*** (▲) *[Prop. 5]*.

Thus, ***DFG*** (▲) is greater than ***EGF*** (▲).

Thus, ***EFG*** (▲+▲) is much greater than ***EGF*** (▲). And since triangle ***EFG*** has angle ***EFG*** (▲+▲) greater than ***EGF*** (▲), and the greater angle is subtended by the greater side *[Prop. 19]*.

Side ***EG*** (—) is thus also greater than ***EF*** (—). But ***EG*** (—) is equal to ***BC*** (—). Thus, ***BC*** (—) is also greater than ***EF*** (—).

Thus, if two triangles have two sides equal to two sides, respectively, but one has the angle encompassed by the equal straight-lines greater than the corresponding angle in the other, then the former triangle will also have a base greater than the base of the latter

Q.E.D

Proposition 25

If two triangles have two sides equal to two sides, respectively, but (one) has a base greater than the base of the other, then the former triangle will also have the angle encompassed by the equal straight-lines greater than the corresponding angle in the latter.

Note: This proves the converse to the last proposition. That is, given two triangles with different sized bases, the triangle with the wider base will have the angle at its tip wider than the tip of the other triangle. The proof is not as elegant as the previous proposition, but that is just personal preference. Proposition 25 considers the three cases, the angle is the same, less or bigger. It eliminates the two cases that the angles are the same or one is less than the other, leaving only the third remaining. It therefore concludes that the angle must be larger. It's a proof by elimination, so what's left must be the truth.

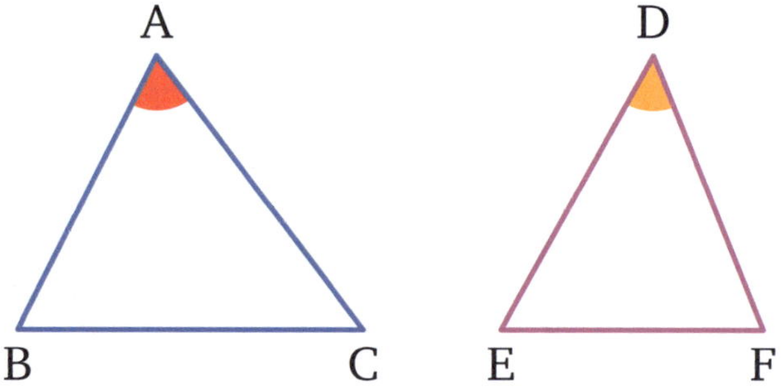

Let ***ABC*** and ***DEF*** be two triangles having the two sides ***AB*** and ***AC*** equal to the two sides ***DE*** and ***DF***, respectively That is, ***AB*** equal to ***DE***, and ***AC*** to ***DF***.

Let the base ***BC*** be greater than the base ***EF***. I say that angle ***BAC*** (∠) is also greater than ***EDF***.

For if not, ***BAC*** (∠) is either equal to, or less than, ***EDF*** (∠). In fact, ***BAC*** (∠) is not equal to ***EDF*** (∠). For then the base ***BC*** would also have been equal to the base ***EF*** *[Prop. 4]*.

But it is not. Thus, angle ***BAC*** (∠) is not equal to ***EDF*** (∠). Neither, indeed, is ***BAC*** (∠) less than ***EDF*** (∠). For then the base ***BC*** would also have been less than the base ***EF*** *[Prop. 24]*.

But it is not. Thus, angle ***BAC*** (∠) is not less than ***EDF*** (∠). But it was shown that (***BAC***, ∠) is not equal to ***EDF*** (∠) either. Thus, ***BAC*** (∠) is greater than ***EDF*** (∠).

Thus, if two triangles have two sides equal to two sides, respectively, but one has a base greater than the base of the other, then the former triangle will also have the angle encompassed by the equal straight-lines greater than the corresponding angle in the latter. Q.E.D

Proposition 26

If two triangles have two angles equal to two angles, respectively, and one side equal to one side—in fact, either that by the equal angles, or that subtending one of the equal angles—then the triangles will also have the remaining sides equal to the [corresponding] remaining sides, and the remaining angle equal)to the remaining angle.

Note: This is the third and last triangle congruence proof in Euclid's elements. This one proves the ASA principle, that is angle-side-angle. The previous congruence theorems were SAS and SSS. Unlike the previous two proofs, Proposition 26 doesn't use the superposition concept and in that sense Proposition 26 is on more solid foundations. The proof itself is quite long as it considers two cases and in each case uses contradiction as the method of proof.

Let ***ABC*** (▲) and ***DEF*** (▲) be two triangles having the two angles ***ABC*** (∠) and ***BCA*** (∠) equal to the two angles ***DEF*** (∠) and ***EFD*** (∠), respectively. That is ***ABC*** (∠) equal to ***DEF*** (∠), and ***BCA*** (∠) to ***EFD*** (∠). And let them also have one side equal to one side. First of all, the side by the equal angles.

That is ***BC*** equal to ***EF***. I say that they will have the remaining sides equal to the corresponding remaining sides. That is ***AB*** equal to ***DE***, and ***AC*** to ***DF***.

And they will have the remaining angle equal to the remaining angle. That is ***BAC*** (∠) equal to ***EDF*** (∠).

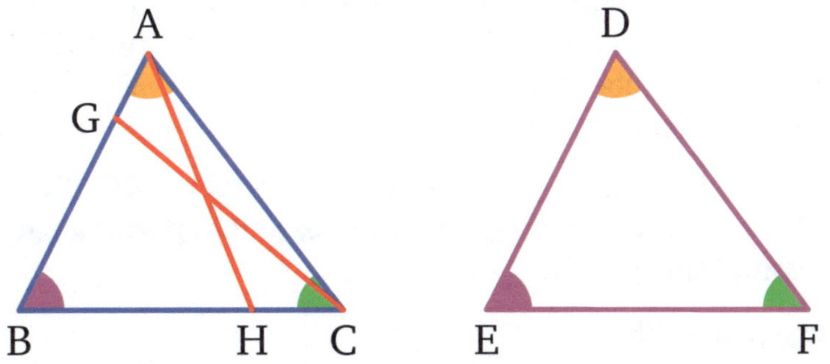

For if ***AB*** is unequal to ***DE*** then one of them is greater. Let ***AB*** be greater, and let ***BG*** be made equal to ***DE*** *[Prop. 3]*,

and let ***GC*** have been joined.

Therefore, since ***BG*** is equal to ***DE***, and ***BC*** to ***EF***, the two straight-lines ***GB***, ***BC*** are equal to the two straight-lines ***DE***, ***EF***, respectively. And angle ***GBC*** is equal to angle ***DEF*** (△). Thus, the base ***GC*** is equal to the base ***DF***, and triangle ***GBC*** (△) is equal to triangle ***DEF*** (△) , and the remaining angles subtended by the equal sides will be equal to the corresponding remaining angles *[Prop. 4]*.

Thus, ***GCB*** is equal to ***DFE*** (△). But, ***DFE*** (△) was assumed to be equal to ***BCA*** (△). Thus, ***BCG*** is also equal to ***BCA*** (△), the lesser to the greater. The very thing is impossible. Thus, ***AB*** is not unequal to ***DE***. Thus, it is equal.

Also ***BC*** is equal to ***EF***. So the two straight-lines ***AB***, ***BC*** are equal to the two straight-lines ***DE***, ***EF***, respectively.

And angle ***ABC*** (△) is equal to angle ***DEF*** (△). Thus, the base ***AC*** is equal to the base ***DF***, and the remaining angle ***BAC*** (△) is equal to the remaining angle ***EDF*** (△) *[Prop. 4]*.

But, again, let the sides subtending the equal angles be equal: for instance, let ***AB*** be equal to ***DE***. Again, I say that the remaining sides will be equal to the remaining sides. That is ***AC*** equal to ***DF***, and ***BC*** to ***EF***. Furthermore, the remaining angle ***BAC*** (△) is equal to the remaining angle ***EDF*** (△).

For if ***BC*** is unequal to ***EF*** then one of them is greater. If possible, let ***BC*** be greater. And let ***BH*** be made equal to ***EF*** *[Prop. 3]*,

Let ***AH*** have been joined.

And since ***BH*** is equal to ***EF***, and ***AB*** to ***DE***, the two straight-lines ***AB***, ***BH*** are equal to the two straight-lines ***DE***, ***EF***, respectively. And the angles they encompass (are also equal). Thus, the base ***AH*** is equal to the base ***DF***, and the triangle ***ABH*** is equal to the triangle ***DEF***, and the remaining angles subtended by the equal sides will be equal to the corresponding remaining angles *[Prop. 4]*.

Thus, angle ***BHA*** is equal to ***EFD*** (△). But, ***EFD*** (△) is equal to ***BCA*** (△). So, in triangle ***AHC***, the external angle ***BHA*** is equal to the internal and opposite angle ***BCA*** (△). The very thing is impossible *[Prop. 16]*.

Thus, ***BC*** is not unequal to ***EF***.

Thus, it is equal. And ***AB*** is also equal to ***DE***. So the two straight-lines ***AB***, ***BC*** are equal to the two straight-lines ***DE***, ***EF***, respectively. And they encompass equal angles. Thus, the base ***AC*** is equal to the base ***DF***, and triangle ***ABC*** is equal to triangle ***DEF***, and the remaining angle ***BAC*** (△) is equal to the remaining angle ***EDF*** (△) *[Prop. 4]*.

Thus, if two triangles have two angles equal to two angles, respectively, and one side equal to one side—in fact, either that by the equal angles, or that subtending one of the equal angles—then the triangles will also have the remaining sides equal to the corresponding remaining sides, and the remaining angle equal to the remaining angle.

Q.E.D

Proposition 27

If a straight-line falling across two straight-lines makes the alternate angles equal to one another then the two straight-lines will be parallel to one another.

Note: At this point Euclid switches to a new topic, leaving behind proofs that relate to triangles and changes to questions concerning parallel lines and parallelograms. The first proof in this series uses contradiction to make its case that if the two interior (∠) but alternate angles are equal, then the two lines are parallel.

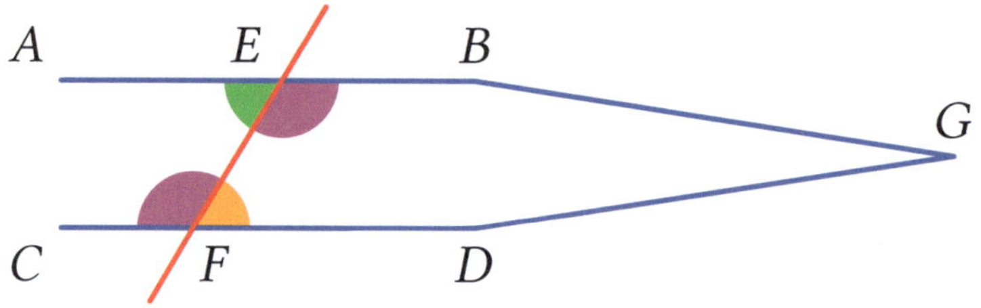

For let the straight-line ***EF*** (—), falling across the two straight-lines ***AB*** (—) and ***CD*** (—), make the alternate angles ***AEF*** (∠) and ***EFD*** (∠) equal to one another. I say that ***AB*** (—) and ***CD*** (—) are parallel.

For if not, being produced, ***AB*** (—) and ***CD*** (—) will certainly meet together: either in the direction of ***B*** and ***D***, or in the direction of ***A*** and ***C*** ***[Def. 23]***.

Let them have been produced, and let them meet together in the direction of ***B*** and ***D*** at point ***G***. So, for the triangle ***GEF*** (∠), the external angle ***AEF*** (∠) is equal to the interior and opposite angle ***EFG*** (∠). The very thing is impossible ***[Prop. 16]***.

Thus, being produced, ***AB*** (—) and ***CD*** (—) will not meet together in the direction of ***B*** and ***D***. Similarly, it can be shown that neither will they meet together in the direction of ***A*** and ***C***. But straight-lines meeting in neither direction are parallel ***[Def. 23]***.

Thus, ***AB*** (—) and ***CD*** (—) are parallel.

Thus, if a straight-line falling across two straight-lines makes the alternate angles equal to one another then the two straight-lines will be parallel to one another. Q.E.D

Proposition 28

If a straight-line falling across two straight-lines makes the external angle equal to the internal and opposite angle on the same side, or makes the sum of the internal angles on the same side equal to two right-angles, then the two straight-lines will be parallel to one another.

> **Note:** The previous proposition showed that if the two interior but adjacent angles are the same, the two lines were parallel. This proposition goes further and proves two additional patterns in the angles. The first is that if the interior ∠ and exterior ∠ angles are equal, then the lines are parallel. Secondly, if the sum of the two inner angles is 180°, then the two lines are parallel.

Let ***EF*** (—), falling across the two straight-lines ***AB*** (—) and ***CD*** (—), make the external angle ***EGB*** (∠) equal to the internal and opposite angle ***GHD*** (∠), or the sum of the internal angles on the same side, ***BGH*** (∠) and ***GHD*** (∠), equal to two right-angles. I say that ***AB*** (—) is parallel to ***CD*** (—).

For since in the first case ***EGB*** (∠) is equal to ***GHD*** (∠), but ***EGB*** (∠) is equal to ***AGH*** (∠) *[Prop. 15]*.

AGH (∠) is thus also equal to ***GHD*** (∠). And they are alternate angles. Thus, ***AB*** (—) is parallel to ***CD*** (—) *[Prop. 27]*.

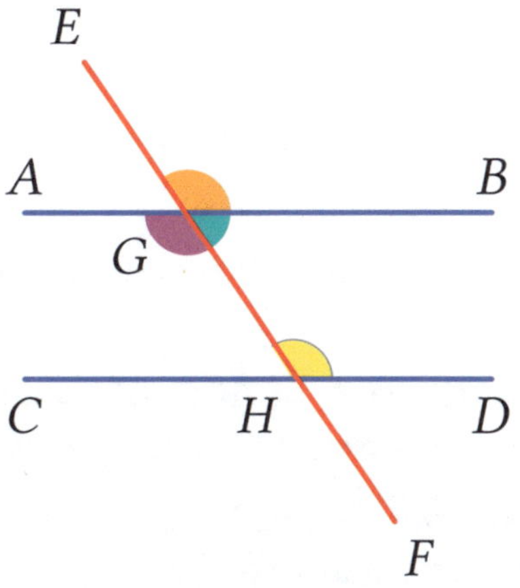

Since in the second case, the sum of ***BGH*** (∠) and ***GHD*** (∠) is equal to two right-angles, and the sum of ***AGH*** (∠) and ***BGH*** (∠) is also equal to two right-angles *[Prop. 13]*,

the sum of ***AGH*** (∠) and ***BGH*** (∠) is thus equal to the sum of ***BGH*** (∠) and ***GHD*** (∠).

Let ***BGH*** (∠) have been subtracted from both. Thus, the remainder ***AGH*** (∠) is equal to the remainder ***GHD*** (∠).

And they are alternate angles. Thus, ***AB*** (—) is parallel to ***CD*** (—) *[Prop. 27]*.

Thus, if a straight-line falling across two straight-lines makes the external angle equal to the internal and opposite angle on the same side, or makes the sum of the internal angles on the same side equal to two right-angles, then the two straight-lines will be parallel to one another.

Q.E.D

Proposition 29

A straight-line falling across parallel straight-lines makes the alternate angles equal to one another, the external angle equal to the internal and opposite angle, and the sum of the internal angles on the same side equal to two right-angles.

> **Note:** This proposition simply proves the converse of the previous proposition. The wording on this proposition might require some clarification due to the way it's written. It's simply proposing that ∠ = ∠, that ∠ = ∠, and that ∠ + ∠ = 180°.

For let the straight-line ***EF*** (—) fall across the parallel straight-lines ***AB*** (—) and ***CD*** (—).

I say that it makes the alternate angles, ***AGH*** (∠) and ***GHD*** (∠), equal, the external angle ***EGB*** (∠) equal to the internal and opposite angle ***GHD*** (∠), and the sum of the internal angles on the same side, ***BGH*** (∠) and ***GHD*** (∠), equal to two right-angles.

For if ***AGH*** (∠) is unequal to ***GHD*** (∠) then one of them is greater.

Let ***AGH*** (∠) be greater.

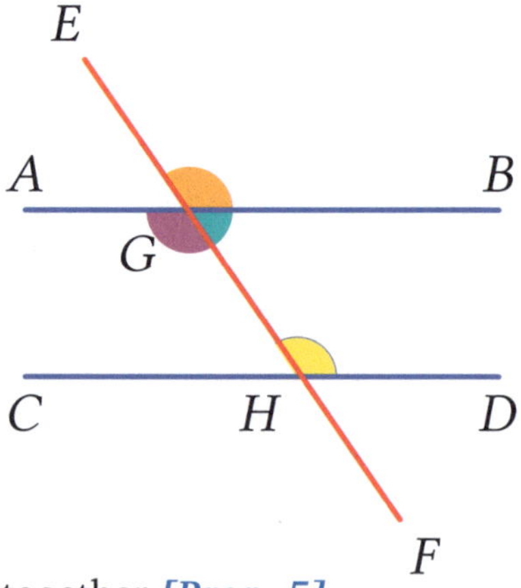

Let ***BGH*** (∠) have been added to both.

Thus, the sum of ***AGH*** (∠) and ***BGH*** (∠) is greater than the sum of ***BGH*** (∠) and ***GHD*** (∠).

But, the sum of ***AGH*** (∠) and ***BGH*** (∠) is equal to two right-angles *[Prop. 13]*.

Thus, the sum of ***BGH*** (∠) and ***GHD*** (∠) is also less than two right-angles. But straight-lines being produced to infinity from internal angles whose sum is less than two right-angles meet together *[Prop. 5]*.

Thus, ***AB*** and ***CD***, being produced to infinity, will meet together. But they do not meet, on account of them initially being assumed parallel to one another ***[Def. 23]***.

Thus, ***AGH*** (∠) is not unequal to ***GHD*** (∠). Thus, it is equal. But, ***AGH*** (∠) is equal to ***EGB*** (∠) ***[Prop. 15]***.

And ***EGB*** (∠) is thus also equal to ***GHD*** (∠).

Let ***BGH*** (∠) be added to both. Thus, the sum of ***EGB*** (∠) and ***BGH*** (∠) is equal to the sum of ***BGH*** (∠) and ***GHD*** (∠).

But, the sum of ***EGB*** (∠) and ***BGH*** (∠) is equal to two right-angles ***[Prop. 13]***.

Thus, the sum of ***BGH*** (∠) and ***GHD*** (∠) is also equal to two right-angles.

Thus, a straight-line falling across parallel straight-lines makes the alternate angles equal to one another, the external angle equal to the internal and opposite angle, and the sum of the internal angles on the same side equal to two right-angles.

Q.E.D

Proposition 30

Straight-lines parallel to the same straight-line are also parallel to one another.

> **Note:** This is reminiscent of Axiom 1 which states that if $a = c$ and $b = c$, then $a = b$. In this case a, b, and c are parallel lines.

Let each of the straight-lines ***AB*** (——) and ***CD*** (——) be parallel to ***EF*** (——). I say that ***AB*** (——) is also parallel to ***CD*** (——).

For let the straight-line ***GK*** (——) fall across ***AB*** (——), ***CD*** (——), and ***EF*** (——).

And since the straight-line ***GK***(——) has fallen across the parallel straight-lines ***AB*** (——) and ***EF*** (——), angle ***AGK*** (∠) is thus equal to ***GHF*** (∠) *[Prop. 29]*.

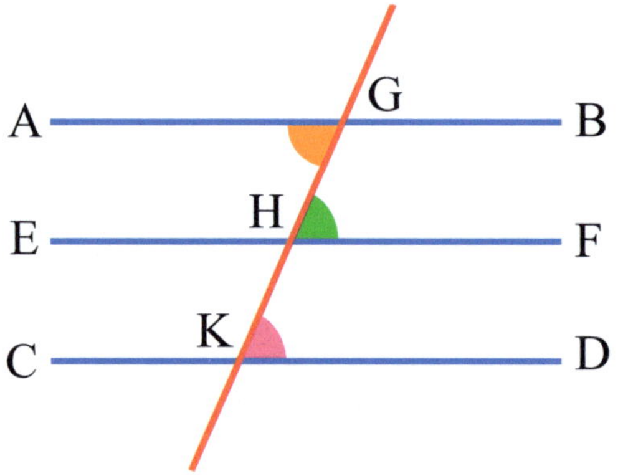

Again, since the straight-line ***GK*** (——) has fallen across the parallel straight-lines ***EF*** (——) and ***CD*** (——), angle ***GHF*** (∠) is equal to ***GKD*** (∠) *[Prop. 29]*.

But ***AGK*** (∠) was also shown to be equal to ***GHF*** (∠). Thus, ***AGK*** (∠) is also equal to ***GKD*** (∠). And they are alternate angles. Thus, ***AB*** (——) is parallel to ***CD*** (——) *[Prop. 27]*.

Q.E.D

Proposition 31

To draw a straight-line parallel to a given straight-line, through a given point.

This is a critically important proposition that is used in nine subsequent propositions. It's a construction on how to draw a line parallel to another line. The most common modern construction technique uses a different approach to Proposition 31 which requires a compass-distance measure that would have been much less appealing to the Greeks.

Let ***A*** be the given point, and ***BC*** (——) the given straight-line. It is required to draw through point ***A*** a straight-line (———) parallel to the straight-line ***BC*** (——).

Let the point ***D*** have been taken a random on ***BC*** (——), and let ***AD*** (——) have been joined.

Let let angle ***DAE*** (∠), equal to angle ***ADC*** (∠), have been constructed on the straight-line ***DA*** (——) at the point ***A*** on it *[Prop. 23]*.

And let the straight-line ***AF*** (——) have been produced in a straight-line with ***EA*** (——).

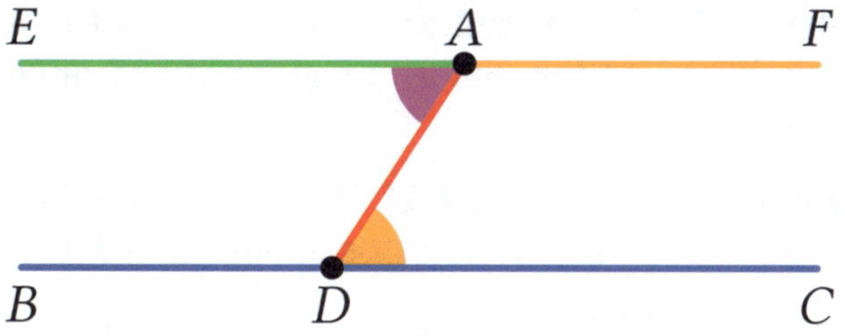

And since the straight-line ***AD*** (——), in falling across the two straight-lines ***BC*** (——) and ***EF*** (———), has made the alternate angles ***EAD*** (∠) and ***ADC*** (∠) equal to one another, ***EAF*** is thus parallel to ***BC*** (——) *[Prop. 27]*.

Thus, the straight-line ***EAF*** (———) has been drawn parallel to the given straight-line ***BC*** (——), through the given point ***A***.

Q.E.F

Proposition 32

In any triangle, if one of the side is produced then the external angle is equal to the sum of the two internal and opposite angles, and the sum of the three internal angles of the triangle is equal to two right-angles.

Note: This is the proposition where Euclid shows that the sum of the three interior angles in a triangle add up to 180°. Something every high School student should know. This is not the proof that tends to be used today, however. The modern proof is called the Pythagorean proof and is perhaps simpler to teach. Both proofs are old and predate Euclid according to Proclus, however the one we use today probably came later, but for whatever reason Euclid kept the earlier one.

Let ***ABC*** be a triangle, and let one of its sides ***BC*** (—) have been produced to ***D***. I say that the external angle ***ACD*** (∠ + ∠) is equal to the sum of the two internal and opposite angles ***CAB*** (∠) and ***ABC*** (∠), and the sum of the three internal angles of the triangle ***ABC*** (∠), ***BCA*** (∠), and ***CAB*** (∠), is equal to two right-angles.

Let ***CE*** (—) have been drawn through point ***C*** parallel to the straight-line ***AB*** (—) *[Prop. 31]*. And since ***AB*** (—) is parallel to ***CE*** (—), and ***AC*** (—) has fallen across them, the alternate angles ***BAC*** (∠) and ***ACE*** (∠) are equal to one another *[Prop. 29]*.

Since ***AB*** (∠) is parallel to ***CE*** (—), and the straight-line ***BD*** (—) has fallen across them, the external angle ***ECD*** (∠) is equal to the internal and opposite angle ***ABC*** (∠) *[Prop. 29]*.

But ***ACE*** (∠) was also shown to be equal to ***BAC*** (∠). Thus, the whole angle ***ACD*** (∠+∠) is equal to the sum of the two internal and opposite angles ***BAC*** (∠) and ***ABC*** (∠).

Let ***ACB*** (∠) have been added to both. Thus, the sum of ***ACD*** (∠ + ∠) and ***ACB*** (∠) is equal to the sum of the three angles ***ABC*** (∠), ***BCA*** (∠), and ***CAB*** (∠). But, the sum of ***ACD*** (∠+∠) and ***ACB*** (∠) is equal to two right-angles *[Prop. 13]*. Thus, the sum of ***ACB*** (∠), ***CBA*** (∠), and ***CAB*** (∠) is also equal to two right-angles.

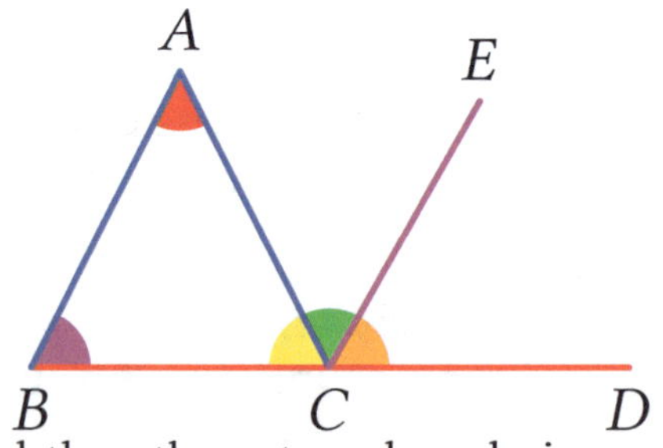

Thus, in any triangle, if one of the sides is produced then the external angle is equal to the sum of the two internal and opposite angles, and the sum of the three internal angles of the triangle is equal to two right-angles. Q.E.D

Proposition 33

Straight-lines joining equal and parallel straight-lines on the same sides are themselves also equal and parallel.

Note: Euclid now switches to a new topic, the properties of parallelograms. This requires the previous results on triangles and parallel lines.

Let ***AB*** (—) and ***CD*** (—) be equal and parallel straight-lines, and let the straight-lines ***AC*** (—) and ***BD*** (—) join them on the same sides. I say that ***AC*** (—) and ***BD*** (—) are also equal and parallel.

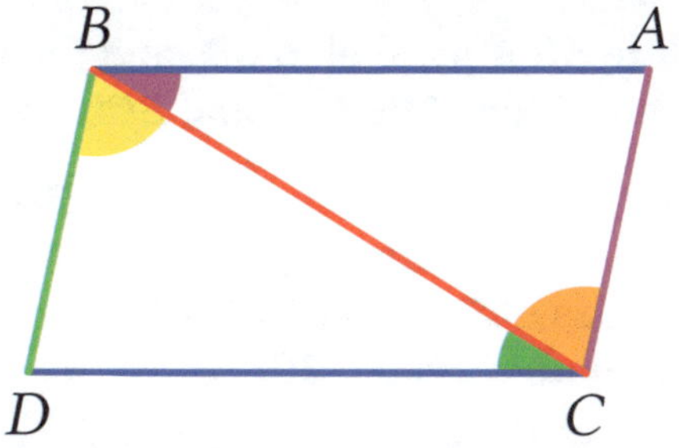

Let ***BC*** (—) have been joined. And since ***AB*** (—) is parallel to ***CD*** (—), and ***BC*** (—) has fallen across them, the alternate angles ***ABC*** (∠) and ***BCD*** (∠) are equal to one another ***[Prop. 29]***.

And since ***AB*** (—) is equal to ***CD*** (—), and ***BC*** (—) is common, the two straight-lines ***AB*** (—), ***BC*** (—) are equal to the two straight-lines ***DC*** (—), ***CB*** (—).

And the angle ***ABC*** (∠) is equal to the angle ***BCD*** (∠). Thus, the base ***AC*** (—) is equal to the base ***BD*** (—), and triangle ***ABC*** (———) is equal to triangle ***DCB*** (———), and the remaining angles will be equal to the corresponding remaining angles subtended by the equal sides ***[Prop. 4]***.

Thus, angle ***ACB*** (∠) is equal to ***CBD*** (∠). Also, since the straight-line ***BC*** (—), in falling across the two straight-lines ***AC*** (—) and ***BD*** (—), has made the alternate angles ***ACB*** (∠) and ***CBD*** (∠) equal to one another, ***AC*** (—) is thus parallel to ***BD*** (—) ***[Prop. 27]***.

And ***AC*** (—) was also shown to be equal to ***BD*** (—).

Thus, straight-lines joining equal and parallel straight-lines on the same sides are themselves also equal and parallel. Q.E.D

Proposition 34

In parallelogrammic figures the opposite sides and angles are equal to one another, and a diagonal cuts them in half.

Note: That is: ∡ = ∡ and ∡ + ∡ = ∡ + ∡. When Euclid says 'cuts them in half he means the area is cut into two equal halves.

Let ***ACDB*** be a parallelogrammic figure, and ***BC*** (—) its diagonal. I say that for parallelogram ***ACDB***, the opposite sides and angles are equal to one another, and the diagonal ***BC*** (—) cuts it in half.

For since ***AB*** is parallel to ***CD***, and the straight-line ***BC*** (—) has fallen across them, the alternate angles ***ABC*** (∡) and ***BCD*** (∡) are equal to one another ***[Prop. 29]***.

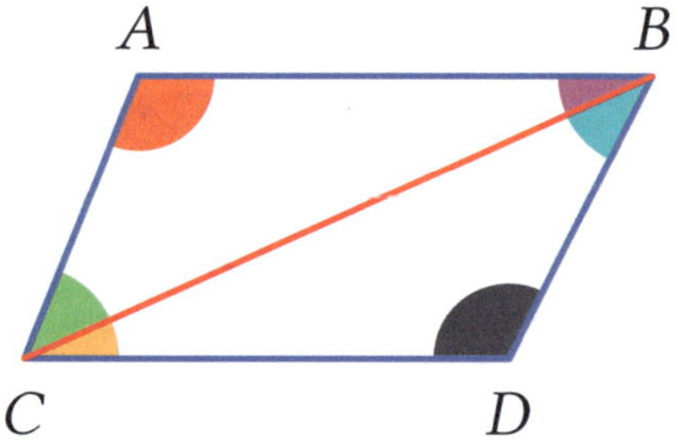

Again, since ***AC*** is parallel to ***BD***, and ***BC*** (—) has fallen across them, the alternate angles ***ACB*** (∡) and ***CBD*** (∡) are equal to one another ***[Prop. 29]***.

So ***ABC*** and ***BCD*** are two triangles having the two angles ***ABC*** (∡) and ***BCA*** (∡) equal to the two angles ***BCD*** (∡) and ***CBD*** (∡), respectively, and one side equal to one side, namely that adjoining the equal angles and common to both of them, namely ***BC*** (—). Thus, they will also have the remaining sides equal to the corresponding remaining sides, and the remaining angle equal to the remaining angle ***[Prop. 26]***.

Thus, side ***AB*** is equal to ***CD***, and ***AC*** to ***BD***. Furthermore, angle ***BAC*** (∡) is equal to ***CDB*** (∡). And since angle ***ABC*** (∡) is equal to ***BCD*** (∡), and ***CBD*** (∡) to ***ACB*** (∡), the whole angle ***ABD*** (∡+∡) is thus equal to the whole angle ***ACD*** (∡+∡).

And ***BAC*** (∡) was also shown to be equal to ***CDB*** (∡). Thus, in parallelogrammic figures the opposite sides and angles are equal to one another.

And, I also say that a diagonal cuts them in half. For since ***AB*** is equal to ***CD***, and ***BC*** (—) is common, the two straight-lines ***AB***, ***BC*** (—) are equal to the two straight-lines ***DC***, ***CB*** (——), respectively. And angle ***ABC*** (△) is equal to angle ***BCD*** (△). Thus, the base ***AC*** is also equal to ***DB***, and triangle ***ABC*** is equal to triangle ***BCD*** *[Prop. 4]*.

Thus, the diagonal ***BC*** (——) cuts the parallelogram ***ACDB*** in half. Q.E.D

Proposition 35

Parallelograms which are on the same base and between the same parallels are equal to one another.

Note: It may not be completely obvious, but when Euclid refers to things being equal in this proof, he actually means equal areas. For example when he says 'and triangle EAB will be equal to triangle DFC' he means the areas of the two triangles are equal. The same applies to the other shapes he refers to.

The proposition shows that the two parallelograms ***ABCE*** (■) and ***EBCF*** (■) have the same area. It's a clever proof involving identifying two similar triangles using SAS, then spotting equal sub-areas of the parallelogram using subtraction and finally adding in a common area to show the parallelograms have the same area. It's an interesting proof employing multiple applications of geometric ideas. One thing the figure doesn't show well are the two embedded triangles ***EAB*** and ***DFC***, which is why I added a second figure on the right to help the reader. These are critical to the proof.

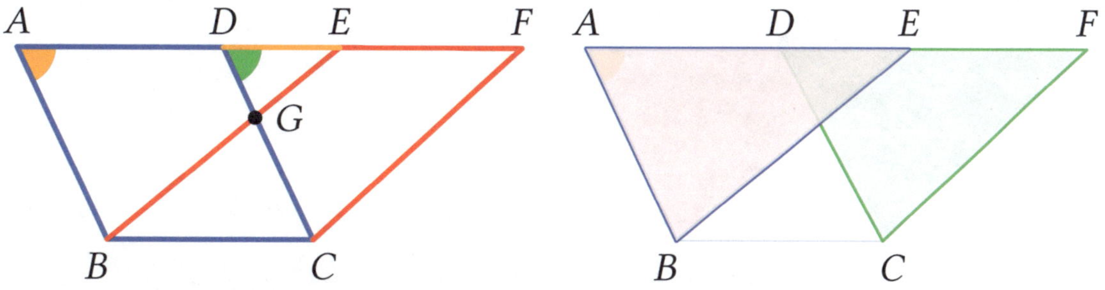

Let ***ADCB*** (■) and ***EBCF*** (■) be parallelograms on the same base ***BC*** (——), and between the same parallels ***AF*** (————) and ***BC*** (——). I say that ***ADCB*** (■) is equal to parallelogram ***EBCF*** (■).

For since ***ADCB*** (■) is a parallelogram, ***AD*** (——) is equal to ***BC*** (——) *[Prop. 34]*.

So, for the same reasons, ***EF*** (——) is also equal to ***BC*** (——). So ***AD*** (——) is also equal to ***EF*** (—). And ***DE*** (—) is common.

Thus, the whole straight-line ***AE*** (——) is equal to the whole straight-line ***DF*** (——).

And ***AB*** (—) is also equal to ***DC*** (—).

So the two straight-lines ***EA*** (——), ***AB*** (—) are equal to the two straight-lines ***FD*** (——), ***DC*** (—), respectively.

And angle ***FDC*** (∠) is equal to angle ***EAB*** (∠), the external to the internal ***[Prop. 29]***.

Thus, the base ***EB*** (—) is equal to the base ***FC*** (—), and triangle ***EAB*** (△) will be equal to triangle ***DFC*** (△) ***[Prop. 4]***.

Let ***DGE*** have been taken away from both.

Thus, the remaining trapezium ***ABGD*** is equal to the remaining trapezium ***EGCF***. Let triangle ***GBC*** have been added to both. Thus, the whole parallelogram ***ADCB*** (■) is equal to the whole parallelogram ***EBCF*** (■).

Thus, parallelograms which are on the same base and between the same parallels are equal to one another.

Q.E.D

Proposition 36

Parallelograms which are on equal bases and between the same parallels are equal to one another.

This is a generalization of the previous proposition. Just imagine the red parallelogram in 34 moved to the right and you've got this proposition.

Let ***ABCD*** (■) and ***EFGH*** (■) be parallelograms which are on the equal bases ***BC*** (——) and ***FG*** (——), and are between the same parallels ***AH*** (————) and ***BG*** (————). I say that the parallelogram ***ABCD*** (■) is equal to ***EFGH*** (■).

Join ***BE*** (——) and ***CH*** (——).

Since ***BC*** (——) is equal to ***FG*** (——), but ***FG*** (——) is equal to ***EH*** (——) *[Prop. 34]*,

BC (——) is thus equal to ***EH*** (——). And they are also parallel, and ***EB*** (——) and ***HC*** (——) join them.

But straight-lines joining equal and parallel straight-lines on the same sides are themselves equal and parallel *[Prop. 33]*.

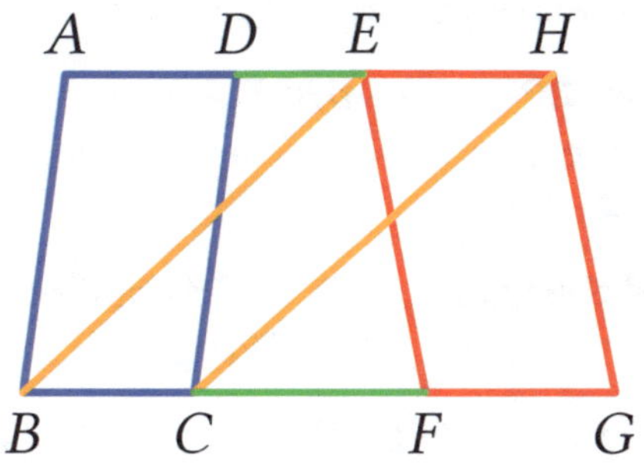

Thus, ***EBCH*** (■) is a parallelogram *[Prop. 34]*, and is equal to ***ABCD*** (■). For it has the same base, ***BC*** (——), as ***ABCD*** (■), and is between the same parallels, ***BC*** (——) and ***AH*** (————), as ***ABCD*** (■) *[Prop. 35]*.

So, for the same reasons, ***EFGH*** (■) is also equal to the same parallelogram ***EBCH*** (■) *[Prop. 34]*.

So that the parallelogram ***ABCD*** (■) is also equal to ***EFGH*** (■).

Thus, parallelograms which are on equal bases and between the same parallels are equal to one another. Q.E.D

Proposition 37

Triangles which are on the same base and between the same parallels are equal to one another.

Note: The following two propositions, 37 and 38, are similar to the previous two except these deal with areas of triangles instead of parallelograms.

Let ***ABC*** (▲) and ***DBC*** (▲) be triangles on the same base ***BC*** (——), and between the same parallels ***AD*** (——) and ***BC*** (——). I say that triangle ***ABC*** (▲) is equal to triangle ***DBC*** (▲).

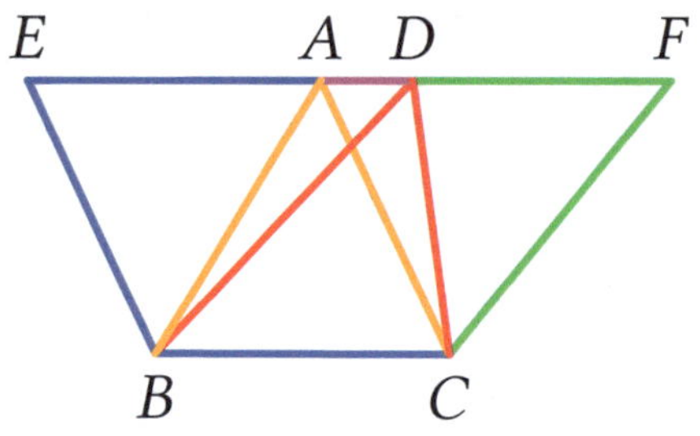

Let ***AD*** (——) have been produced in both directions to ***E*** and ***F***, and let the straight-line ***BE*** (——) have been drawn through ***B*** parallel to ***CA*** (——) *[Prop. 31]*,

Let the straight-line ***CF*** (——) have been drawn through ***C*** parallel to ***BD*** (——) *[Prop. 31]*.

Thus, ***EBCA*** and ***DBCF*** are both parallelograms, and are equal. For they are on the same base ***BC*** (——), and between the same parallels ***BC*** (——) and ***EF*** (————) *[Prop. 35]*.

And the triangle ***ABC*** (▲) is half of the parallelogram ***EBCA***. For the diagonal ***AB*** (——) cuts the latter in half *[Prop. 34]*.

And the triangle ***DBC*** (▲) is half of the parallelogram ***DBCF***. For the diagonal ***DC*** (——) cuts the latter in half *[Prop. 34]*.

Thus, triangle ***ABC*** (▲) is equal to triangle ***DBC*** (▲).

Thus, triangles which are on the same base and between the same parallels are equal to one another.

Q.E.D

Proposition 38

Triangles which are on equal bases and between the same parallels are equal to one another.

Note: Compare this to Proposition 36. Just as Proposition 36 is a generalization of Proposition 35, this is a generalization of Proposition 37.

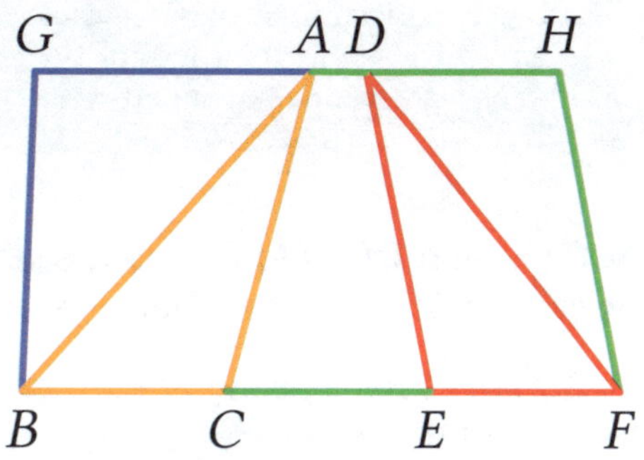

Let ***ABC*** (▲) and ***DEF*** (▲) be triangles on the equal bases ***BC*** (——) and ***EF*** (——), and between the same parallels ***BF*** (————) and ***AD*** (——). I say that triangle ***ABC*** (▲) is equal to triangle ***DEF*** (▲).

For let ***AD*** (——) have been produced in both directions to ***G*** and ***H***, and let the straight-line (——) have been drawn through ***B*** parallel to ***CA*** (——) *[Prop. 31]*, and let the straight-line ***FH*** (——) have been drawn through ***F*** parallel to ***DE*** (——) *[Prop. 31]*.

Thus, ***GBCA*** and ***DEFH*** are each parallelograms. And ***GBCA*** is equal to ***DEFH***. For they are on the equal bases ***BC*** (——) and ***EF*** (——), and between the same parallels ***BF*** (————) and ***GH*** (————) *[Prop. 36]*.

And triangle ***ABC*** (▲) is half of the parallelogram ***GBCA***. For the diagonal ***AB*** (——) cuts the latter in half *[Prop. 34]*.

And triangle ***FED*** (▲) is half of parallelogram ***DEFH***. For the diagonal ***DF*** (——) cuts the latter in half. Thus, triangle ***ABC*** (▲) is equal to triangle ***DEF*** (▲).

Thus, triangles which are on equal bases and between the same parallels are equal to one another.

Q.E.D

Proposition 39

Equal triangles which are on the same base, and on the same side, are also between the same parallels.

Note: By "the same side" Euclid means the two triangles are on the same side of the base line. In the figure they are indicted by ***ABC*** (▲) and ***DBC*** (▲). In some respects this is an odd proposition because since the triangles are the same they must be the same height, and since they are on the same base, a line drawn across the upper tips of the triangles must be parallel to the base line. The proof is however interesting, it assumes that the drawn parallel in the figure, ***AD*** (——), is not parallel and instead suggests that the line ***AE*** (——) is the actual parallel line. By doing this however, we end up with a contradiction, as Euclid is oft to state "The very thing is impossible". This means that the original line ***AD*** (——) was in fact the true parallel line.

Let ***ABC*** (▲) and ***DBC*** (▲) be equal triangles which are on the same base ***BC*** (——), and on the same side of it. I say that they are also between the same parallels.

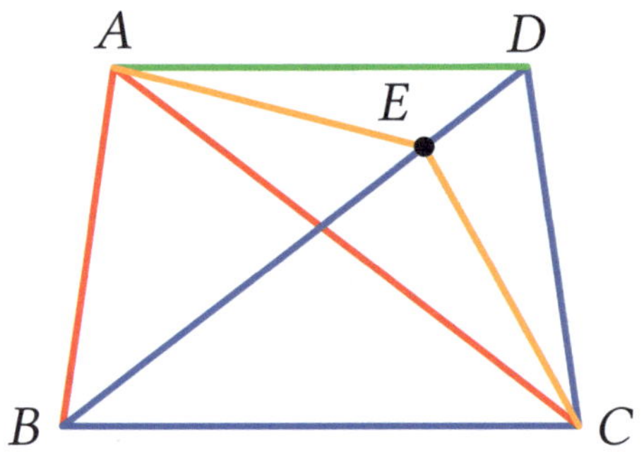

For let ***AD*** (——) have been joined. I say that ***AD*** (——) and ***BC*** (——) are parallel.

For, if not, let ***AE*** (——) have been drawn through point ***A*** parallel to the straight-line ***BC*** (——) *[Prop. 31]*,

and let ***EC*** (——) have been joined. Thus, triangle ***ABC*** (▲) is equal to triangle ***EBC***. For it is on the same base as it, ***BC*** (——), and between the same parallels *[Prop. 37]*.

But ***ABC*** (▲) is equal to ***DBC*** (▲). Thus, ***DBC*** (▲) is also equal to ***EBC***, the greater to the lesser. The very thing is impossible.

Thus, ***AE*** (——) is not parallel to ***BC*** (——). Similarly, we can show that neither is any other straight-line than ***AD*** (——). Thus, ***AD*** (——) is parallel to ***BC*** (——).

Thus, equal triangles which are on the same base, and on the same side, are also between the same parallels. Q.E.D

Proposition 40

Equal triangles which are on equal bases, and on the same side, are also between the same parallels.

> This can be thought of as a generalization of the last proposition. There is evidence to suggest that this proposition was added at a later date and doesn't adhere to the usual standards for a Euclidian proposition. Moreover, the proposition is not used anywhere in all 13 books. It was likely added to match the previous related pairs of propositions, namely 35 to 36 and 37 to 38.

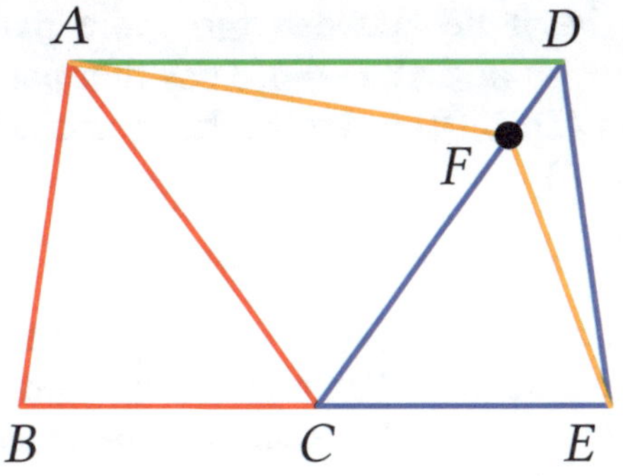

Let ***ABC*** (▲) and ***CDE*** (▲) be equal triangles on the equal bases ***BC*** (——) and ***CE*** (——) respectively, and on the same side of ***BE*** (————). I say that they are also between the same parallels.

For let ***AD*** (——) have been joined. I say that ***AD*** (——) is parallel to ***BE*** (————).

For if not, let ***AF*** (——) have been drawn through ***A*** parallel to ***BE*** (————) ***[Prop. 31]***,

and let ***FE*** (——) have been joined. Thus, triangle ***ABC*** (▲) is equal to triangle ***FCE***. For they are on equal bases, ***BC*** (——) and ***CE*** (——), and between the same parallels, ***BE*** (————) and ***AF*** (——) ***[Prop. 38]***.

But, triangle ***ABC*** (▲) is equal to triangle ***DCE*** (▲). Thus, triangle ***DCE*** (▲) is also equal to triangle ***FCE***, the greater to the lesser. The very thing is impossible.

Thus, ***AF*** (——) is not parallel to ***BE*** (————). Similarly, we can show that neither is any other straight-line than ***AD*** (——). Thus, ***AD*** (——) is parallel to ***BE*** (————).

Thus, equal triangles which are on equal bases, and on the same side, are also between the same parallels.

Q.E.D

Proposition 41

If a parallelogram has the same base as a triangle, and is between the same parallels, then the parallelogram is double the area of the triangle.

Note: Now a slight change of topic. This proposition is the first to give a quantitative result related to areas, namely that the area of a parallelogram is twice the area of a triangle. The proof is surprisingly simple.

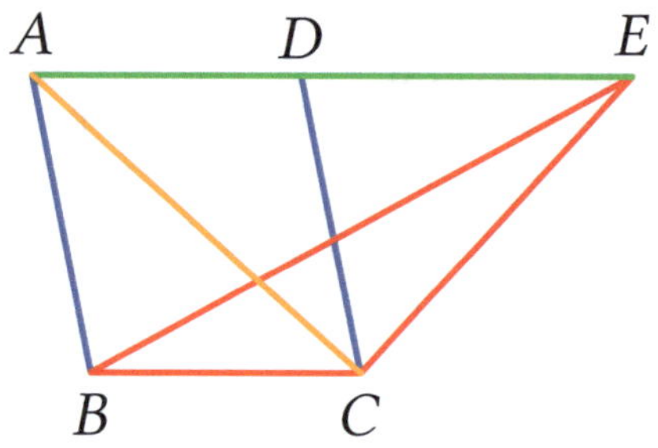

For let parallelogram ***ADCB*** (■) have the same base ***BC*** (—) as triangle ***EBC*** (▲), and let it be between the same parallels, ***BC*** (—) and ***AE*** (—). I say that parallelogram ***ADCB*** (■) is double the area of triangle ***BEC*** (▲).

For let ***AC*** (—) have been joined.

So triangle ***ABC*** (———) is equal to triangle ***EBC*** (▲). For it is on the same base, ***BC*** (—), as ***EBC*** (▲), and between the same parallels, ***BC*** (—) and ***AE*** (—) *[Prop. 37]*.

But, parallelogram ***ADCB*** (■) is double the area of triangle ***ABC*** (———). For the diagonal ***AC*** (—) cuts the former in half *[Prop. 34]*.

So parallelogram ***ADCB*** (■) is also double the area of triangle ***EBC*** (▲).

Thus, if a parallelogram has the same base as a triangle, and is between the same parallels, then the parallelogram is double the area of the triangle.

Q.E.D

Proposition 42

Construct a parallelogram equal to a given triangle in a given rectilinear angle.

> This is a clever construction, and it might take a few readings to grasp its logic, but as you can see, it uses five previous propositions.

Let ***ABC*** (▲) be the given triangle, and ***D*** the given rectilinear angle. So it is required to construct a parallelogram equal to triangle ***ABC*** (▲) in the rectilinear angle ***D*** (∠).

Let ***BC*** (—) have been cut in half at ***E*** *[Prop. 10]*,

and let ***AE*** (—) have been joined. And let angle ***CEF*** (∠), equal to angle ***D*** (∠), have been constructed at the point ***E*** on the straight-line ***EC*** (—) *[Prop. 23]*.

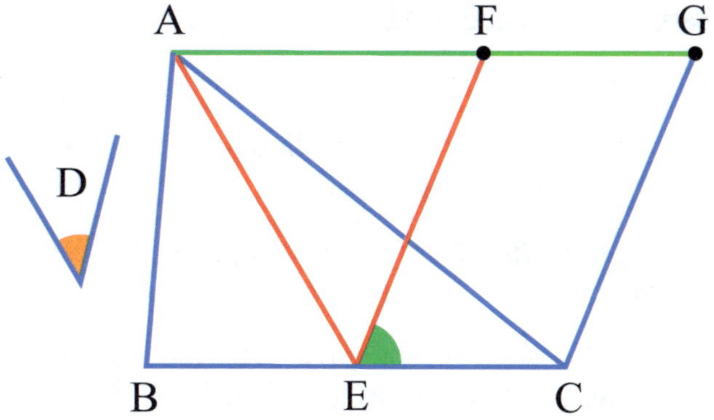

And let ***AG*** (—) have been drawn through ***A*** parallel to ***EC*** (—) *[Prop. 31]*,

and let ***CG*** (—) have been drawn through ***C*** parallel to ***EF*** (—) *[Prop. 31]*.

Thus, ***FECG*** is a parallelogram. And since ***BE*** (—) is equal to ***EC*** (—), triangle ***ABE*** is also equal to triangle ***AEC***. For they are on the equal bases, ***BE*** (—) and ***EC*** (—), and between the same parallels, ***BC*** (—) and ***AG*** (—) *[Prop. 38]*.

Thus, triangle ***ABC*** (▲) is double the area of triangle ***AEC***. And parallelogram ***FECG*** is also double the area of triangle ***AEC***. For it has the same base as ***AEC***, and is between the same parallels as ***AEC*** *[Prop. 41]*.

Thus, parallelogram ***FECG*** is equal to triangle ***ABC*** (▲). ***FECG*** also has the angle ***CEF*** (∠) equal to the given angle ***D***.

Thus, parallelogram ***FECG***, equal to the given triangle ***ABC*** (▲), has been constructed in the angle ***CEF*** (∠), which is equal to ***D*** (∠). Q.E.F

Proposition 43

For any parallelogram, the complements of the parallelograms about the diagonal are equal to one another.

> This proposition is much simpler that it seems. The large parallelogram, ***ABCD***, is split into two halves by ***AC*** (——). Each half has two triangles, △ and △. The triangles have the same area in both halves. That leaves the complements with the same area. Algebraically if we designate the two half areas as T, and the two triangles as T_1 and T_2, and the complements as C_1 and C_2, we would write: $T = T_1 + T_2 + C_1$ and $T = T_1 + T_2 + C_2$, setting them equal to each other and canceling T_1 and T_2 we are left with $C_1 = C_2$.

Let ***ABCD*** (■) be a parallelogram, and ***AC*** (——) its diagonal. And let ***EH*** (□) and ***FG*** (□) be the parallelograms about ***AC*** (——), and ***BK*** (□) and ***KD*** (□) the so-called complements about ***AC***.

I say that the complement ***BK*** (□) is equal to the complement ***KD*** (□).

For since ***ABCD*** (■) is a parallelogram, and ***AC*** (——) its diagonal, triangle ***ABC*** is equal to triangle ***ACD*** *[Prop. 34]*.

Again, since ***EH*** (□) is a parallelogram, and ***AK*** (——) is its diagonal, triangle ***AEK*** (△) is equal to triangle ***AHK*** (△) *[Prop. 34]*.

So, for the same reasons, triangle ***KFC*** (△) is also equal to triangle ***KGC***. Therefore, since triangle ***AEK*** (△) is equal to triangle ***AHK*** (△), and ***KFC*** (△) to ***KGC*** (△), triangle ***AEK*** (△) plus ***KGC*** (△) is equal to triangle ***AHK*** (△) plus ***KFC*** (△). And the whole triangle ***ABC*** is also equal to the whole triangle ***ADC***. Thus, the remaining complement ***BK*** (□) is equal to the remaining complement ***KD*** (□).

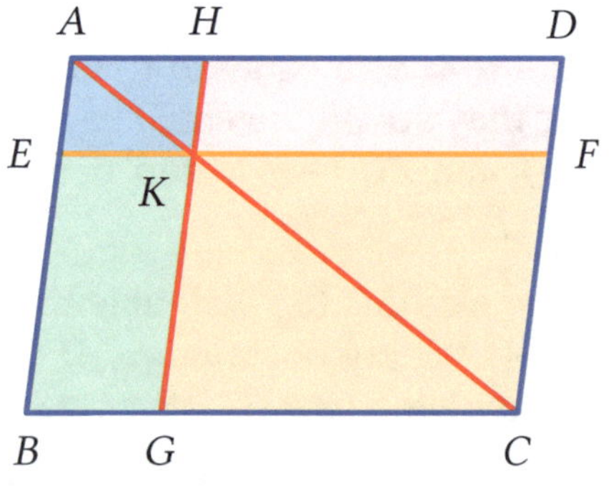

Thus, for any parallelogramic figure, the complements of the parallelograms about the diagonal are equal to one another.

Q.E.D

Proposition 44

To apply a parallelogram equal to a given triangle to a given straight-line in a given rectilinear angle.

Note: This is a construction that creates a parallelogram, ***LB*** (□), with an area given by a triangle, ***C*** (▲), where the corner angle, ***ABM***, is a given angle ***D*** (∠) (=***EBG***). There are two steps in this construction. The first constructs a parallelogram with the correct angle and equal in area to the given triangle. The second changes its length to the proper length, ***AB*** (——), while keeping the same area. Heath comments that this is the most impressive proposition in all geometry. Proclus gives even more praise by calling those who devised this construction as "godlike men of old".

Let ***AB*** (——) be the given straight-line, ***C*** (▲) the given triangle, and ***D*** (∠) the given rectilinear angle. So it is required to apply a parallelogram equal to the given triangle ***C*** (▲) to the given straight-line ***AB*** (——) in an angle equal to angle ***D*** (∠).

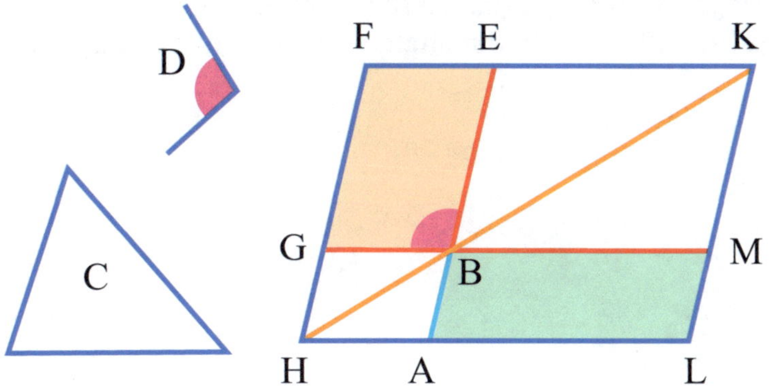

Let the parallelogram ***BEFG*** (□), equal to the triangle ***C*** (▲), have been constructed in the angle ***EBG*** (∠), which is equal to ***D*** (∠) *[Prop. 42]*.

And let it have been placed so that ***BE*** (——) is straight-on to ***AB*** (——). And let ***FG*** (——) have been drawn through to ***H***, and let ***AH*** have been drawn through ***A*** parallel to either of ***BG*** or ***EF*** *[Prop. 31]*,

and let ***HB*** (——) have been joined. And since the straight-line ***HF*** (——) falls across the parallels ***AH*** and ***EF***, the sum of the angles ***AHF*** and ***HFE*** is thus equal to two right-angles *[Prop. 29]*.

Thus, the sum of ***BHG*** and ***GFE*** is less than two right-angles.

And straight-lines produced to infinity from internal angles whose sum is less than two right-angles meet together *[Post. 5]*.

Thus, being produced, ***HB*** (——) and ***FE*** (——) will meet together. Let them have been produced, and let them meet together at ***K***. And let ***KL*** have been drawn through point ***K*** parallel to either of ***EA*** or ***FH*** *[Prop. 31]*.

And let ***HA*** and ***GB*** have been produced to points ***L*** and ***M*** respectively. Thus, ***HLKF*** is a parallelogram, and ***HK*** (——) its diagonal.

And ***AG*** and ***ME*** are parallelograms, and ***LB*** (■) and ***BF*** (■) the so-called complements, about ***HK*** (——). Thus, ***LB*** (■) is equal to ***BF*** (■) *[Prop. 43]*.

But, ***BF*** (■) is equal to triangle ***C*** (▲). Thus, ***LB*** (□) is also equal to ***C*** (▲). Also, since angle ***GBE*** (∠) is equal to ***ABM*** *[Prop. 15]*,

but ***GBE*** (∠) is equal to ***D*** (∠), ***ABM*** is thus also equal to angle ***D*** (∠).

Thus, the parallelogram ***LB*** (■), equal to the given triangle ***C*** (▲), has been applied to the given straight-line ***AB*** (——) in the angle ***ABM***, which is equal to ***D*** (∠).

Q.E.F

Proposition 45

To construct a parallelogram equal to a given rectilinear figure in a given rectilinear angle.

This construction generalizes the previous one and is one of the most important results in Book I. We are no longer limited to converting a triangle into a parallelogram, but any rectilineal shape. As Heath points out, this is a significant result because it represents an example of geometrical algebra.

For example, lets assume that the angle ***E*** is a right-angle, this will lead to a creation of a rectangle with area equal to the rectilinear figure. The area of a rectangle is of course width times height: $w \times h$. Since the area of the rectilineal shape (Area_r) is equal to the area of the rectangle, we can write $\text{Area}_r = w \times h$. Or $w = \text{Area}_r / h$. In other words, the construction is effectively doing a division on two given things, the area of the rectilinear figure and the height of the given line ***AB***. The idea of using geometry as a means for doing algebra is an ingenious one, even if ultimately, too cumbersome. The method is much more developed in Book II, but it has limits, partly because it is too cumbersome and is limited to polynomial equations with rational coefficients. It ultimately stalled further Greek development in mathematics, and it took another 1000 years or more to develop symbolic algebra, which was far easier to use and could deal with a wider range of problems.

There is a historical note from Proclus who suggested that it was this proposition that led the early mathematicians to investigate whether the area of a circle could be copied to a rectangle. This is often shortened to the expression 'squaring the circle'. In 1882, squaring the circle was proven to be impossible using just a compass and rule[a].

It's interesting to note that in Figure 6, Proposition 45 is also the pinnacle in the dependencies chart.

[a] `https://en.wikipedia.org/wiki/Squaring_the_circle`

Let ***ADCB*** (■) be the given rectilinear figure, and ***E*** (▲) the given rectilinear angle. So it is required to construct a parallelogram equal to the rectilinear figure ***ADCB*** (■) in the given angle ***E*** (▲).[19]

Let ***DB*** (—) be joined, and let the parallelogram ***FH*** (□), equal to the triangle ***ABD*** (△), have been constructed in the angle ***HKF***, which is equal to ***E*** (▲) *[Prop. 42]*.

And let the parallelogram ***GM*** (□), equal to the triangle ***DBC*** (△), have been applied to the straight-line ***GH*** (—) in the angle ***GHM***, which is equal to ***E*** (▲) *[Prop. 44]*.

[19] The text here is different from Heath and follows Fitzpatrick's translation which is much clearer.

And since angle ***E*** (∠) is equal to each of angles ***HKF*** (∠) and ***GHM*** (∠), angle ***HKF*** (∠) is also equal to ***GHM*** (∠).

Add the angle ***KHG*** (∠) to both ***GHM*** (∠) and ***KHG*** (∠). Therefore the sum of the angles ***FKH*** (∠) and ***KHG*** (∠) equals the sum of the angles ***KHG*** (∠) and ***GHM*** (∠).

But, the sum of ***FKH*** (∠) and ***KHG*** (∠) is equal to two right-angles *[Prop. 29]*.

Thus, the sum of ***KHG*** (∠) and ***GHM*** (∠) is also equal to two right-angles.

So two straight-lines, ***KH*** and ***HM***, not lying on the same side, make adjacent angles with some straight-line ***GH***, at the point ***H*** on it, whose sum is equal to two right-angles.

Thus, ***KH*** is straight-on to ***HM*** *[Prop. 14]*.

And since the straight-line ***HG*** (—) falls across the parallels ***KM*** and ***FG***, the alternate angles ***MHG*** (∠) and ***HGF*** (∠) are equal to one another *[Prop. 29]*.

Let ***HGL*** (∠) have been added to both. Thus, the sum of ***MHG*** (∠) and ***HGL*** (∠) is equal to the sum of ***HGF*** (∠) and ***HGL*** (∠). But, the sum of ***MHG*** (∠) and ***HGL*** (∠) is equal to two right-angles *[Prop. 29]*.

Thus, the sum of ***HGF*** (∠) and ***HGL*** (∠) is also equal to two right-angles. Thus, ***FG*** is straight-on to ***GL*** *[Prop. 14]*.

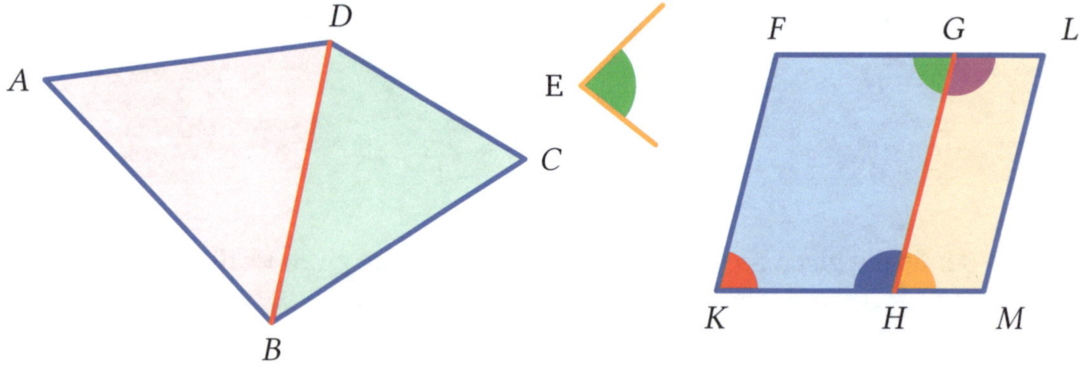

And since ***FK*** is equal and parallel to ***HG*** *[Prop. 34]*,

but also ***HG*** to ***ML*** *[Prop. 34]*, ***KF*** is thus also equal and parallel to ***ML*** *[Prop. 30]*.

And the straight-lines ***KM*** and ***FL*** join them. Thus, ***KM*** and ***FL*** are equal and parallel as well *[Prop. 33]*.

Thus, ***KFLM*** () is a parallelogram. And since triangle ***ABD*** () is equal to parallelogram ***FH*** (), and ***DBC*** () to ***GM*** (), the whole rectilinear figure ***ADCB*** () is thus equal to the whole parallelogram ***KFLM*** ().

Thus, the parallelogram ***KFLM*** (), equal to the given rectilinear figure ***ADCB*** () , has been constructed in the angle ***FKM*** (), which is equal to the given angle ***E***. Q.E.F

Proposition 46

To describe a square on a given straight-line.

Let ***AB*** (——) be the given straight-line. So it is required to describe a square on the straight-line ***AB*** (——).

Let ***AC*** (————) have been drawn at right-angles to the straight-line ***AB*** (——) from the point ***A*** on it ***[Prop. 11]***,

Let ***AD*** (——) have been made equal to ***AB*** (——) ***[Prop. 3]***.

Let ***DE*** (——) have been drawn through point ***D*** parallel to ***AB*** (——) ***[Prop. 31]***,

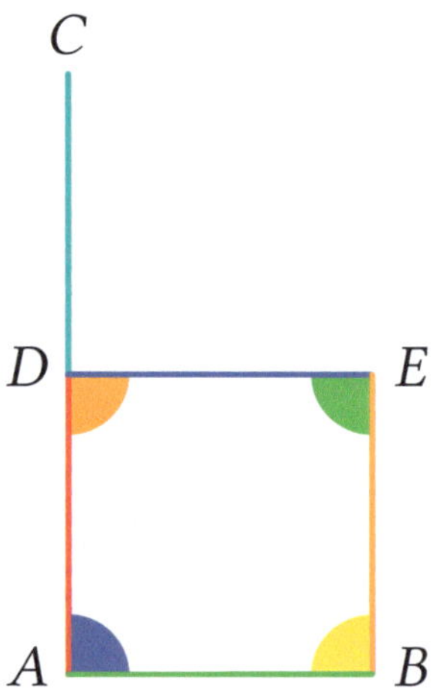

Let ***BE*** (——) have been drawn through point ***B*** parallel to ***AD*** (——) ***[Prop. 31]***.

Thus, ***ADEB*** is a parallelogram. Therefore, ***AB*** (——) is equal to ***DE*** (——), and ***AD*** (——) to ***BE*** (——) ***[Prop. 34]***.

But, ***AB*** (——) is equal to ***AD*** (——). Thus, the four sides ***BA***, ***AD***, ***DE***, and ***EB*** are equal to one another. Thus, the parallelogram ***ADEB*** is equilateral. So I say that it is also right-angled. For since the straight-line ***AD*** (——) falls across the parallels ***AB*** (——) and ***DE*** (——), the sum of the angles ***BAD*** (▲) and ***ADE*** (▲) is equal to two right-angles ***[Prop. 29]***.

But ***BAD*** (▲) is a right-angle. Thus, ***ADE*** (▲) is also a right-angle. And for parallelogrammic figures, the opposite sides and angles are equal to one another ***[Prop. 34]***.

Thus, each of the opposite angles ***ABE*** (▲) and ***BED*** (▲) are also right-angles. Thus, ***ADEB*** is right-angled. And it was also shown to be equilateral.

Thus, ***ADEB*** is a square ***[Def. 22]***.

And it is described on the straight-line ***AB*** (——). Q.E.F

Proposition 47

In right-angled triangles, the square on the side subtending the right-angle is equal to the sum of the squares on the sides containing the right-angle.

We now come to the most famous proposition in Book I, possibly in the whole 13 books of Euclid, popularly known as Pythagoras' Theorem (Proclus). Heath usually gives a page to discuss a given proposition, but for this proposition, he uses 10 pages. Although one can argue that from a theoretical standpoint, Proposition 45 is more important, Proposition 47 has gained infamy while Proposition 45 is hardly known. The reasons are numerous: the first is that Proposition 47 is a result that has direct practical application, secondly the result is accessible to most people who have a high School education, that is $a^2 + b^2 = c^2$, while at the same time remaining somewhat mysterious and thereby intriguing. The proposition has been of so much interest that there are 370 different proofs in the literature.

Let ***ABC*** be a right-angled triangle having the right-angle ***BAC***. I say that the square on ***BC*** is equal to the sum of the squares on ***BA***, ***AC***.

For let there be described on ***BC*** the square ***BDEC*** (□), and on ***BA***, ***AC*** the squares ***GB*** (□), ***HC*** (□) *[Prop. 46]*.

Draw ***AL*** through ***A*** parallel to either ***BD*** or ***CE***, and let ***AD***, and ***FC*** be joined. *[Prop. 31]*.

Since each of the angles ***BAC***, ***BAG*** are right-angles, it follows that with a straight line ***BA***, and at the point ***A*** on it, the two straight lines ***AC***, ***AG***, not lying on the same side, make the adjacent angles equal to two right angles; therefore ***CA*** is in a straight line with ***AG***. *[Prop. 14]*.

For the same reason ***BA*** is also in a straight line with ***AH***.

Since the angle ***DBC*** is equal to the angle ***FBA***, for each is right, let the angle ***ABC*** be added to each; therefore the whole angle ***DBA*** is equal to the whole angle ***FBC*** **[Axiom 2]**.

Since ***DB*** is equal to ***BC***, and ***FB*** to ***BA***, the two sides ***AB***, ***BD*** are equal to the two sides ***FB***, ***BC*** respectively, and the angle ***ABD*** is equal to the angle ***FBC***; therefore the base ***AD*** is equal to the base ***FC***, and the triangle ***ABD*** is equal to the triangle ***FBC*** *[Prop. 4]*.

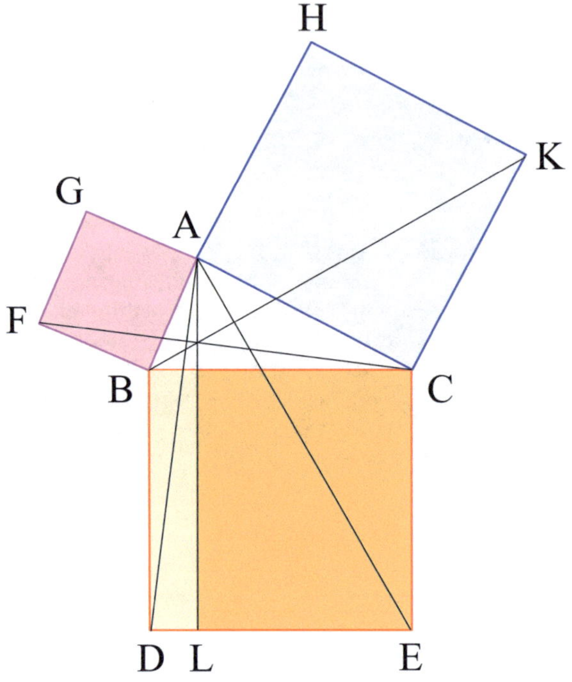

Now the parallelogram ***BL*** is double of the triangle ***ABD***, for they have the same base ***BD*** and are in the same parallels ***BD***, ***AL***.

And the square ***GB*** (□) is double of the triangle ***FBC***, ***[Prop. 41]***, for they again have the same base ***FB*** and are in the same parallels ***FB***, ***GC***.

[But the doubles of equals are equal to one another] Therefore the parallelogram ***BL*** is also equal to the square ***GB***.

Similarly, if ***AE***, ***BK*** are joined, the parallelogram ***CL*** can also be proved equal to the square HC. Therefore the whole square ***BDEC*** (□) is equal to the sum of the two squares ***GB*** (□), ***HC*** (□).

And the square ***BDEC*** (□) is described on ***BC***, and the squares ***GB*** (□), ***HC*** (□) on ***BA***, ***AC***.

Therefore the square on the side ***BC*** is equal to the squares on the sides ***BA***, ***AC***.

Q. E. D.

Proposition 48

If the square on one of the sides of a triangle is equal to the sum of the squares on the two remaining sides of the triangle then the angle contained by the two remaining sides of the triangle is a right-angle.

Note: We come to the final Proposition in Book I. This proposition is just the converse of the Proposition 47 and as such represents a final closing up of Book I.

For let the square on one of the sides, ***BC*** (—), of triangle ***ABC*** be equal to the sum of the squares on the sides ***BA*** (△) and ***AC*** (—). I say that angle ***BAC*** (△) is a right-angle.

For let ***AD*** (—) have been drawn from point ***A*** at right-angles to the straight-line ***AC*** (—) *[Prop. 11]*.

Let ***AD*** (—) have been made equal to ***BA*** (—) *[Prop. 3]*,

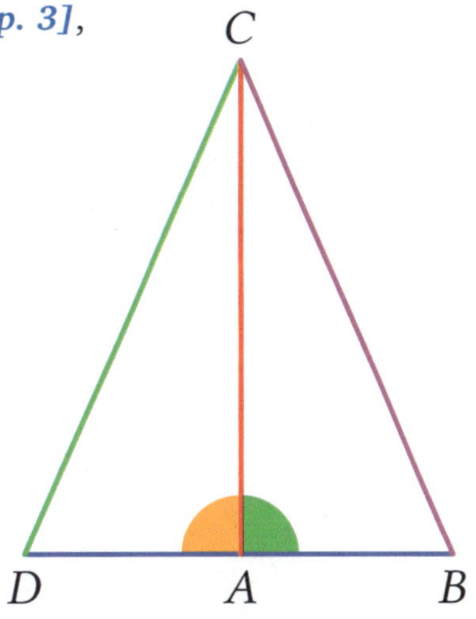

and let ***DC*** (—) have been joined. Since ***DA*** (—) is equal to ***AB*** (—) , the square on ***DA*** (—) is thus also equal to the square on ***AB*** (—). Let the square on ***AC*** (—) have been added to both. Thus, the sum of the squares on ***DA*** (—) and ***AC*** (—) is equal to the sum of the squares on ***BA*** (—) and ***AC*** (—). But, the square on ***DC*** (—) is equal to the sum of the squares on ***DA*** (—) and ***AC*** (—). For angle ***DAC*** (△) is a right-angle *[Prop. 47]*.

But, the square on ***BC*** (—) is equal to sum of the squares on ***BA*** (—) and ***AC*** (—). For that was assumed. Thus, the square on ***DC*** (—) is equal to the square on ***BC*** (—). So side ***DC*** (—) is also equal to side ***BC*** (—).

And since ***DA*** (—) is equal to ***AB*** (—), and ***AC*** (—) is common, the two straight-lines ***DA*** (—), ***AC*** (—) are equal to the two straight-lines ***BA*** (—), ***AC*** (—). And the base ***DC*** (—) is equal to the base ***BC*** (—). Thus, angle ***DAC*** (△) is equal to angle ***BAC*** (△) *[Prop. 8]*.

But ***DAC*** (∠) is a right-angle. Thus, ***BAC*** (∠) is also a right-angle.

Thus, if the square on one of the sides of a triangle is equal to the sum of the squares on the remaining two sides of the triangle then the angle contained by the remaining two sides of the triangle is a right-angle.

Q.E.D

End of Book I

> **Note:** We've come to the end of Book I. It takes effort to get through Euclid, often requiring repeated readings. It's not a perfect book, it has some omissions and undefined elements in places, but even with its faults, it is a remarkable 2300 year old document that provides a systematic and deductive approach to deriving truths about Euclidean space.

8 Appendix

8.1 Proposition 2 and 3

For a first time reader, Proposition 2 and 3 might appear quite puzzling. Geometry books tend to be about proving theorems about triangles, parallel lines etc., not moving lines or cutting off bits of lines which is what Proposition 2 and 3 do. If you want to copy a line, then a natural way is to use a compass to measure the first line, then physically move the compass to the point of interest and draw the new line using the width of the compass. This, for example, is how we draw a line parallel to another line today, by using a compass to measure a specific distance and moving the compass to another location to redraw the line. Proposition 31, which is about drawing parallel lines does not use this modern technique at all. This clearly wasn't something the Greeks wanted to do, which is employ compass-carried distances. Proposition 2 and 3 get around this restriction by using a very clever and purely geometric way to copy a line. The Greeks also wanted to minimise the number of postulates. That is, the smallest number of postulates needed to construct the whole of plane geometry. They could have added an additional postulate that allows lines to be physically moved using a compass, but Proposition 2 allowed them to avoid this.

Copying a line *without* using compass-carried distance is a two part exercise. Proposition 2 first copies the line to a given point, say A. However, it might not be in the right orientation. This is solved by using Proposition 3 which can be used to rotate the line to the desired angle. If you look at Proposition 3, the heart of it is actually quite simple. The real genius is in Proposition 2. It is also quite an involved Proposition, requiring two circles to construct the equilateral triangle, as well as an additional two lines and two circles to complete the construction. The reader starting with the more gentle Proposition 1, is in for a bit of a shock when they get to Proposition 2, especially since it seems completely unnecessary when one could just use a compass to physically measure the length of the line to be copied. However, this is was too crude an operation for the Greeks.

Finally, it is worth mentioning that Proposition 2/3 also brings with it the notion of geometrical arithmetic and even algebra. We should recall that the Greeks, until very late (Diophantus of Alexandria, 200-284 AD), had no modern equivalent of algebra. This meant that all operations had to be carried out within the framework of Geometry, including operations such as addition, subtraction and even multiplication and division. Proposition 2/3 for example, allows lines to be added, subtracted, and lines compared for equality. Given the limitations of Geometry, the Greeks pushed the technique to its limits. It took almost another 1000 years or so to finally move analysis from Geometry to symbolic algebra. Even Newton, as late as the 17th century, presented his proofs in geometric form when he published his Principia. This was due mainly to the fact that geometric proofs were guaranteed to be true by the deductive approach, whereas the 'new' algebraic methods were still immature and open to suspicion. Guicciardini[20] notes that Newton "believed that only geometry could provide a certain and therefore publishable demonstration".

8.2 Proposition 5

Proposition 5 or the Asses' Bridge is a fairly complex proof requiring close study. Historically it has been the bane of students. Interestingly, the Euclid edition by Milliet Dechales from 1685[21], doesn't even give the proof but instead says:

"I was unwilling to make use of Euclid's demonstration, because being very difficult, it might discourage beginners".

Proposition 5 invokes SAS twice on two sets of triangles as well as a subtraction. In this section I will present a more detailed description of the proof and a much simpler proof attributed to Pappus (290-350 AD) by Proclus (410-485 AD) in his commentary on Book I. First let's look at the proposition in more detail. Whether this will enlighten the reader remains to be seen. Euclid's proof, though quite involved, is clever and shows what could be done with minimal information, namely just Propositions 3, and 4 (SAS), with Proposition 4 being the most important. The numbers in the following figures correspond to the numbers in the list of statements that follows.

1) First draw an isosceles triangle with sides of the same length r. The angle at the tip of the triangle is given in purple.

2) Extend the two edges of the triangle to D and E.

We will prove that angles ***ABC*** and ***ACB*** (△) are equal, $\alpha = \beta$ and the angles ***DBC*** and ***ECB*** (▲) are also equal, $\gamma = \delta$.

[20]Guicciardini, Niccol'o, Isaac Newton on Mathematical Certainty and Method, Cambridge Massachusetts: MIT Press, 2009.

[21]"The elements of Euclid explain'd. In a new, but most easie method, together with the use of every proposition through all parts of the mathematicks", Dechales, Claude-Francois Milliet, 1685.

(1)

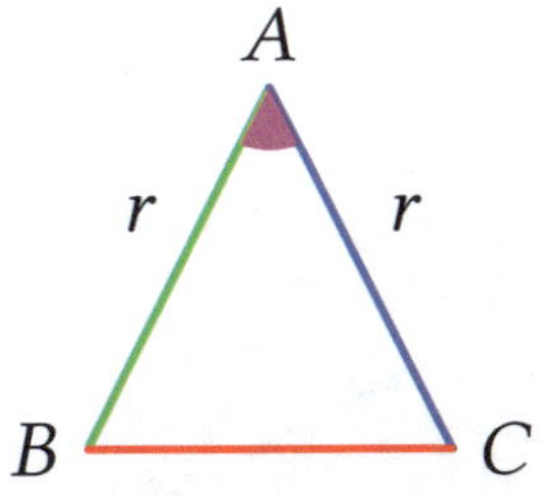

(2)

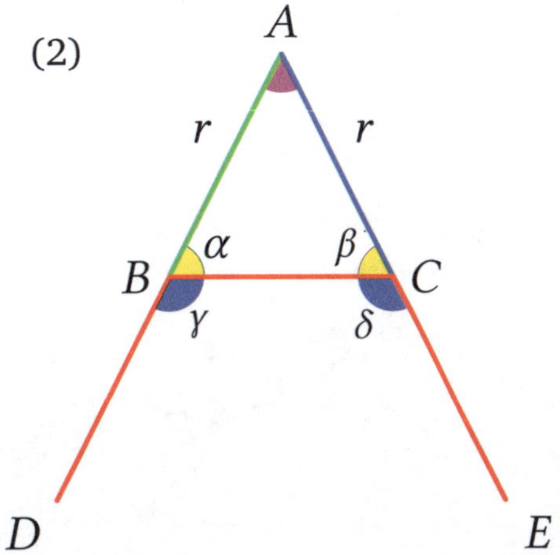

(3) Begin by picking two points F and G on the extended lines where the lengths ***BF*** and ***CG*** are equal. We will designate their length using x.

(4) Let us draw two lines ***FC*** and ***GB*** (——)

(3)

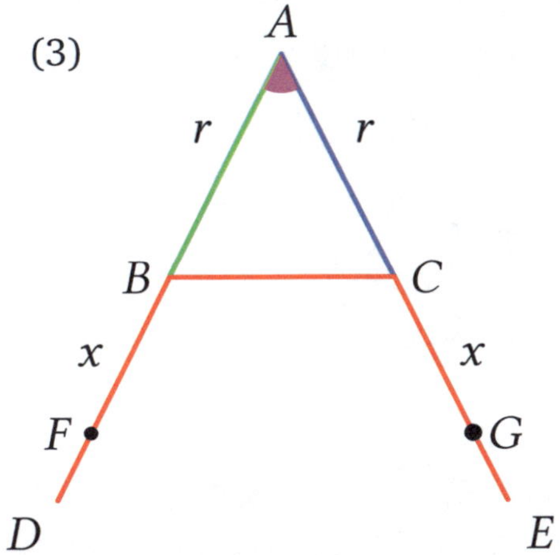

(4)

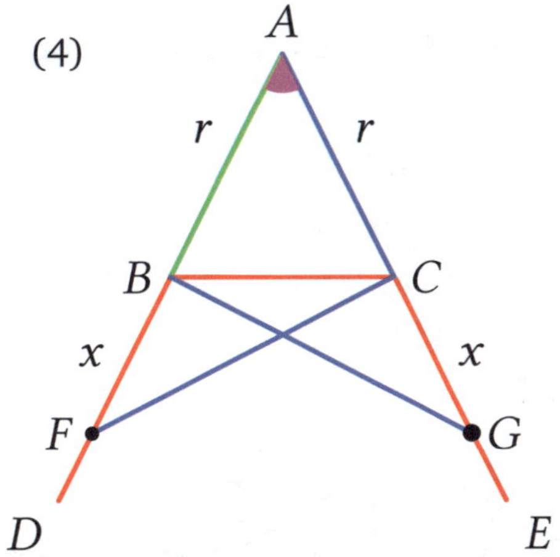

5) If you look carefully you can see two large triangles shown in red (***AGB***) and blue (***AFC***).

(5)

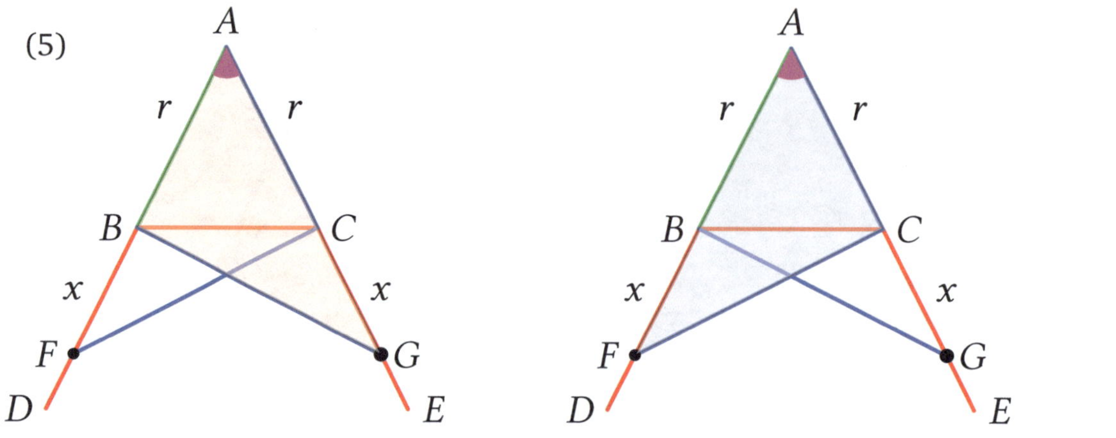

6) Both triangles share the same angle at A (▲). The vertical sides, ABF ($r + x$) and ACG ($r + x$) are the same length. In addition, the two vertical sides AB (—, r) and AC (—, r) are by construction also identical.

By the SAS theorem, we can say that these two triangles are equal or congruent in modern language. If they are equal, then both red angles are equal and both blue angles are equal. In addition, the base lines, y of both triangles are also equal in length.

(6)

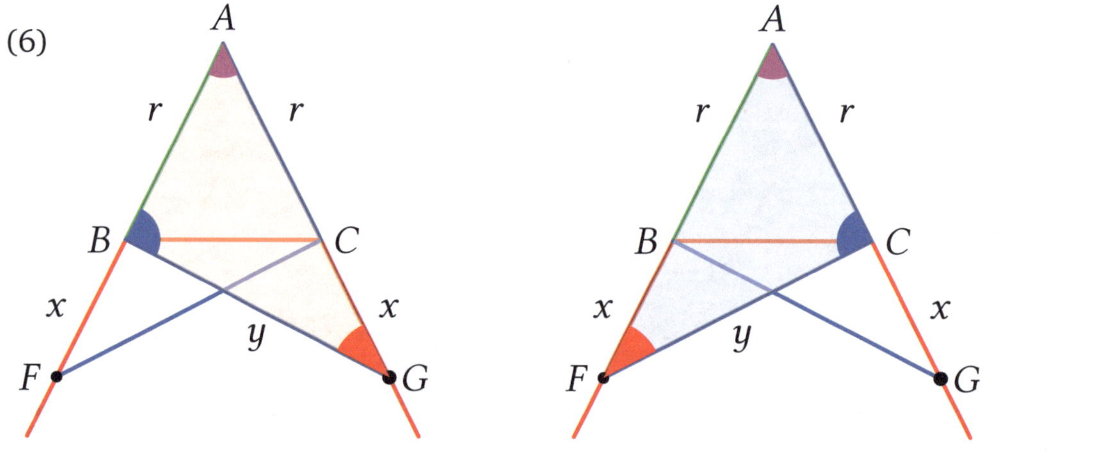

7) The next thing to look at are a new set of triangles in the figure, ***BCF*** (red shading) and ***CBG*** (blue shading). The two triangles have the same angle ***AGB*** (▲) and ***AFC*** (▲) and two sides, x and y; The triangles also share a common side, **BC** (—). Therefore these two triangles are also equal (or congruent). This means that the angles ***CBF*** (▲)

and ***BCG*** (▲) are equal to each other, and the orange angles are equal to each other.

One of the items we wanted to prove was that the angles ***CBF*** (▲) and ***BCG*** (▲) were equal. Hence we have satisfied one part of the proof.

The last thing we must prove is that α and β have the same angles.

(7)

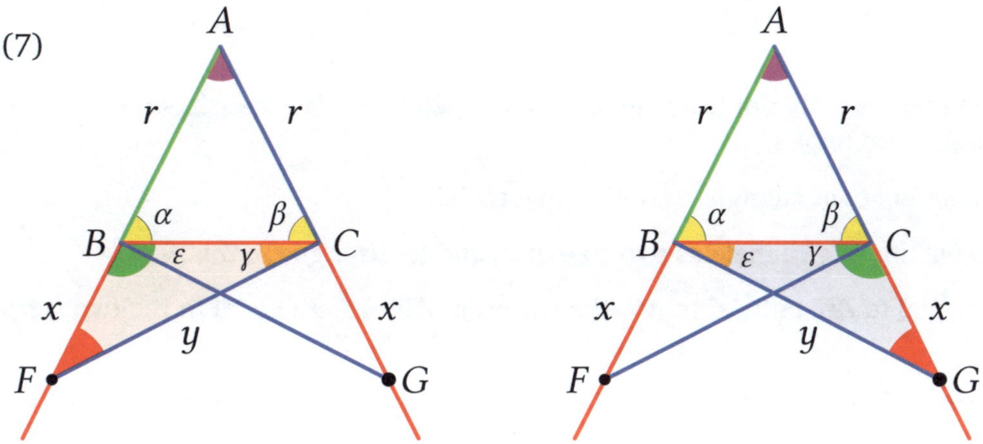

We have already established the following two results:

a). From (7) the two angles ***BCF*** (▲) and ***CBG*** (▲) are equal:

$$\gamma = \varepsilon$$

b). From point (6) we established that the angles ***ABG*** (▲) and ***AFC*** (▲) were the same. However each of these is the sum of the two angles in orange and yellow. We can therefore state:

$$\alpha + \varepsilon = \beta + \gamma$$

But since $\gamma = \varepsilon$, we are left with:

$$\alpha = \beta$$

which is what we set out to demonstrate.

Alternative Proof by Pappus

Pappus considered the following situation. What if we make a copy of the isosceles triangle and spin it around so we see the other side? If we did this we'd get two triangles shown below:

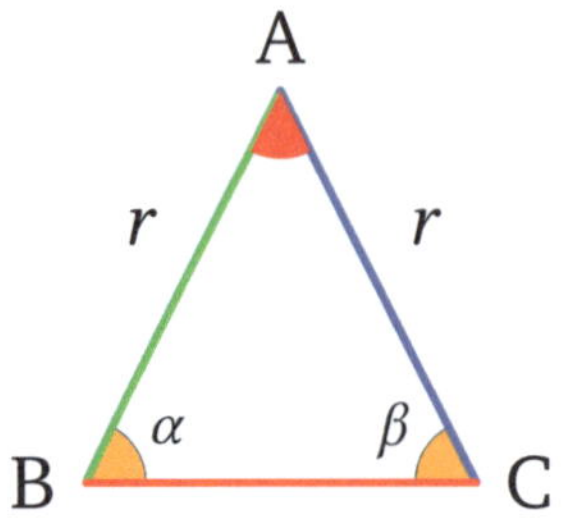

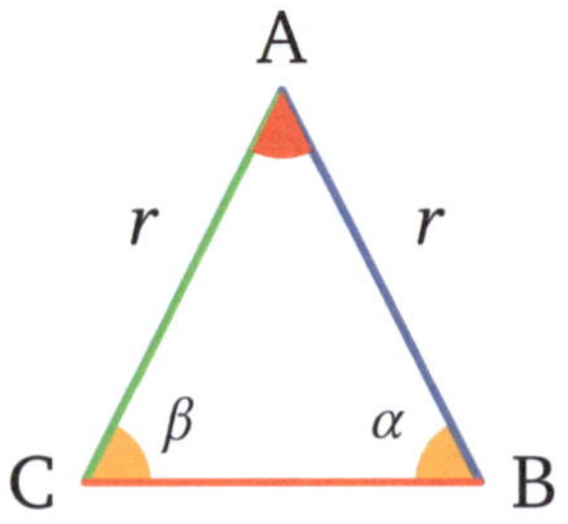

The following text is verbatim from Heath (1956), p254, and is a translation from Pappus' commentary on Book I:

"Let ***ABC*** be an isosceles triangle, and ***AB*** equal to ***AC***.

Let us conceive this one triangle as two triangles, and let us argue in this way.

Since ***AB*** is equal to ***AC***, and ***AC*** to ***AB***, the two sides ***AB***, ***AC*** are equal to the two sides ***AC***, ***AB***.

And the angle ***BAC*** is equal to the angle ***CAB***, for it is the same [i.e they are congruent].

Therefore all the corresponding parts in the triangles are equal, namely

BC to ***BC***

the triangle ***ABC*** to the triangle ***ABC*** (i.e. ***ACB***),

the angle ***ABC*** to the angle ***ACB***,

and the angle ***ACB*** to the angle ***ABC***,

(for these are the angles subtended by the equal sides ***AB***, ***AC***). Therefore in isosceles triangles the angles at the base are equal."

The proof amounts to saying that between the two triangles, $\alpha = \beta$ and $\beta = \alpha$, therefore both base angles are the same.

Modern Proof: From Durell and Arnold, Plane Geometry

Modern proofs use the notation $\angle ABC$ to indicate the angle at ***ABC*** and $\triangle ABC$ to indicate the triangle ***ABC***. I obtained the following proof from Durell and Arnold, Plane Geometry, but it's widely available elsewhere both in books and online. Euclid wasn't able to use this particular proof because it requires the ability to bisect an angle, and that concept wasn't introduced until Proposition 9.

In an isosceles triangle the angles opposite the equal sides are equal.

Given the isosceles $\triangle ABC$ in which $AB = AC$.

To prove $\angle B = \angle C$.

Proof. Let AD be drawn so as to bisect $\angle BAC$.

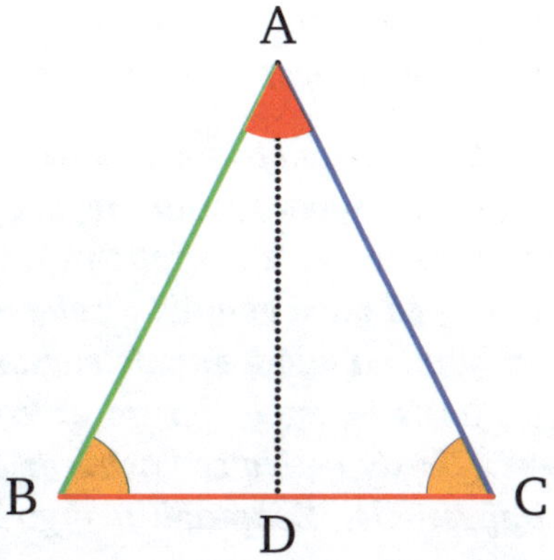

Then, in the triangle ***ABD*** and ***ACD***:

$$AB = AC.$$

Also $$AD = AD,$$

$$\angle BAD = \angle CAD.$$

And $\therefore \triangle ABD = \triangle ACD$.

Two triangles are equal if two sides and the included angle of one are equal, respectively, to two sides and the included angle of the other.

$\therefore \angle B = \angle C$, Q. E. D.

8.3 Proposition Dependencies

The following is a Google Latin translation to English (The translation isn't perfect but good enough for this purpose) of Proposition 4 and 9 from the anonymous Greek to Latin translation that was made in the 12th century at Palermo. It is considered the earliest known translation of its kind (Murdock, 1967). What this manuscript seems to show is that the labels we see today in more modern editions that indicate the dependencies among the propositions, were not apparently included in the very early editions. Labeling appears to be a more modern addition.

Having said that, a papyrus fragment (Oxyrhynchus 29), dated to the turn of first and second centuries CE, may have the number 5 next to the diagram of proposition II.5 (Fowler 1999, p. 210–2 and plate 2 between p. 6 and p. 7), but the evidence is not strong (Saito, 2018).

Proposition 4

Latin:

"Si duo trigona duo latera duobus lateribus equalia habent, utrumque utrique, et angulum angulo equalem habent sub illis equalibus rectis contentum, et basim basi equam habebunt. Et trigonum trigono equale erit et reliqui anguli reliquis angulis equales erunt, uterque utrique, quibus equalia latera subtenduntur.

Sint duo trigona abg et dez, duo latera ab et ag duobus lateribus de et dz equa habentia, utrumque utrique, ab quidem ei quod est de atque ag ei quod est dz et angulum bag angulo edz equalem. Dico quoniam et basis bg basi ez equalis est et abg trigonum dez trigono equale erit. Et reliqui anguli reliquis angulis equales erunt, uterque utrique, quibus equalia latera subtenduntur. Angulus quidem abg angulo dez, angulus vero agb angulo dze.

Coaptato enim abg trigono super dez trigonum et posito a quidem puncto super d punctum, recta vero ab super rectam de, coaptabit et b punctus super punctum e eo quod equalis est recta ab recte de. Coaptata vero recta ab super rectam de, coaptabitur et ag recta super rectam dz eo quod equalis sit bag angulus angulo edz. Quare et g punctus super z punctum coaptabitur eo quod equalis rursus sit recta ag recte dz. Sed et b super e coaptatum erat. Quare basis bg super basim ez coaptabitur. Si enim b quidem super e coaptato, g vero super z, basis bg super basim ez non coaptabitur, due recte spatium continebunt. Quod est impossibile. Coaptabitur ergo basis bg super ez et equalis ipsius erit. Quare et totum abg trigonum super totum dez trigonum coaptabitur et equale ipsius erit. Et reliqui anguli super reliquos angulos coaptabuntur et equales ipsis erunt. Angulus quidem abg angulo dez, angulus vero agb angulo dze. Si ergo duo trigona duo latera duobus lateribus equa habent, utrumque utrique, et angulum angulo equalem habent et que deinceps. Quod oportebat ostendere."

Google Translation:

If two triangles have two sides equal to two sides, both to both, and have an angle equal to the angle contained under those equal lines, and they will have an equal base to the base. And the triangle will be equal to the triangle, and the other angles will be equal to the other angles, each to the two whose equal sides are subtended.

Let there be two triangles *abg* and *dez*, two sides *ab* and *ag* having two sides *de* and *dz* equine, both to both, *ab* indeed to what is *de* and *ag* to it which is *dz* and the angle *b* from *g* is equal to the angle *edz*.

I say that the base *bg* is equal to the base *ez* and *bg* will be equal to the trigon *dez* And the rest of the angles will be equal to the rest of the angles, each to both of which the equal sides are subtended. Indeed, the angle *abg* is at the angle *dez*, but the angle *agb* is at the angle *dze*

For if the triangle *abg* is joined to the triangle *dez* and if the point *a* is indeed placed over the point *d*, and the line *ab* over the line *de*, the point *b* will also join the point *e* over the point *e* by what is equal right *ab* right *de* If the line *ab* is superimposed on the line *de*, the line *ag* will be superimposed on the line *dz* because the angle *bag* is equal to the angle *edz* Therefore, the *g* point will be joined over the *z* point by the fact that the line *ag* is equal to the line *dz*. But *b* was also attached to *e*.

Therefore, the base *bg* will be superimposed on the base *ez*. For if *b* is indeed coagulated over *e*, and *g* is indeed coagulated over *z*, the basis of *bg* will not be coagulated over the base of *ez*, the two will contain a correct space.

That is impossible. Therefore, the base *bg* will be attached to *ez* and will be equal to it. Therefore, the whole *abg* triangle will be superimposed on the whole *dez* triangle and will be equal to it.

And the remaining angles shall be joined upon the remaining angles, and shall be equal to them. Indeed, the angle *abg* is at the angle *dez*, but the angle *agb* is at the angle *dze* If, therefore, two triangles have two sides equal to two sides, each of them has an angle equal to the angle, and so forth. It was necessary to show.

Proposition 9

"Sit datus angulus rectilineus bag. Oportet ergo ipsum in duo equa secare. Sumatur in recta ab punctus quilibet sitque d. Et auferatur a recta ag recte ad equalis recta a e. Et copuletur de. Et constituatur super de trigonum equilaterum dez et copuletur recta az. Dico quoniam bag angulus in duo equa divisus est a recta az.

Quoniam enim equalis est recta ad recte ae, communis vero recta az, due ergo recte da et az duabus rectis ea et az sunt equales, utraque utrique, et basis dz basi ez est equalis, angulus ergo daz angulo eaz equalis est. Datus ergo angulus rectilineus bag in duo equa divisus est a recta az. Quod oportebat ostendere."

Google Translation:

Let the rectilinear angle *bag* be given. It is therefore necessary to cut in two. Let any point be taken on the line *ab* and let *d* be And let the line *ag* be taken from the line *ad* equal to the line a e And *de* will copulate. And let it be established over *de* the equilateral triangle *dez* and the straight line *az* be joined. I say that the angle *bag* is bisected by the line *az*.

For since the straight line *ad* is equal to the straight line *ae*, but the common straight line *az* is equal, therefore due to the straight lines *da* and *az* the two straight lines *ea* and *az* are equal, both to both, and the base *dz* the base *ez* is equal, therefore the angle *d* from *z* is equal to the angle *e* from *z* Given then the rectilinear angle *bag* is bisected by the line *az*. It was necessary to show.

8.4 What a proof might look like today

Although there are legitimate reservations to changing the language used in Euclid's Elements, it is interesting to speculate what a more modern but still readable version might look like. Here is a possible example of Proposition 1 using a more modern style.

Constructing an Equilateral Triangle

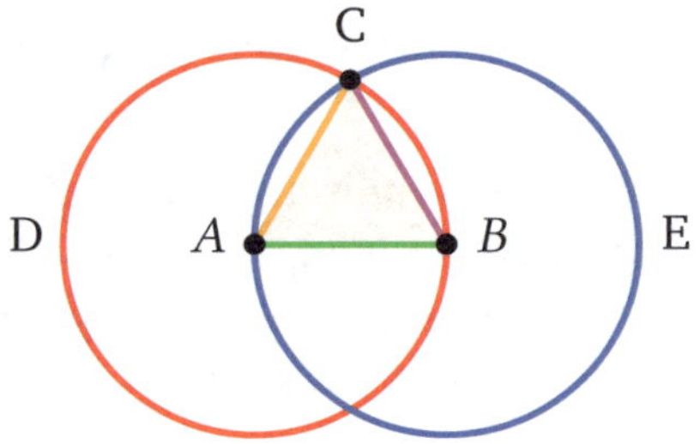

Aim:

Construct an equilateral triangle.

Construction:

1. Draw a straight-line ***AB*** (—).
2. Draw a circle, ***BCD*** (●), with center ***A*** and radius ***AB*** (—).
3. Draw a circle, ***ACE*** (●), with center ***B*** and radius ***BA*** (—).
4. Where the circles cross, mark the point with the letter ***C***.
5. Join the points ***A*** to ***C*** and ***B*** to ***C*** with straight lines forming a triangle ***ABC***.
6. The shape ***ABC*** (———) is an equilateral triangle.

Proof:

1. Since point ***A*** is the center of the circle ***BCD*** (●), ***AC*** (—) is equal to ***AB*** (—).
2. Since point ***B*** is the center of the circle ***ACE*** (●), ***BC*** (—) is equal to ***BA*** (—).
3. That is ***AB*** = ***AC*** or — = —,
4. and ***AB*** = ***BC*** or — = —,
5. Therefore ***AC*** = ***BC*** or — = —. Thus, the three straight lines equal each other

   AB = ***AC*** = ***BC*** or — = — = —

which is the definition of an equilateral triangle.

Revisions

1.0	First public release
1.01 (July, 2023)	page 30, ASS should ASA
1.02 (Nov 5th, 2023)	Updates on Adelard of Bath (p11), minor corrections and clarifications to text, p9, p14, p16, p19, p21, p23, p39, p120

Made in the USA
Monee, IL
05 November 2026

67858051R00070